Metal-Organic Framework Membranes for Molecular Gas Separations

SERIES ON CHEMICAL ENGINEERING

Series Editor: Jerry Y. S. Lin *(Arizona State Univ., USA)*

Vol. 6 *Metal-Organic Framework Membranes for Molecular Gas Separations*
by Moises A. Carreon & Surendar R. Venna

Vol. 5 *Organic-Inorganic Composite Membranes for Molecular Separation*
by Wanqin Jin, Gongping Liu & Nanping Xu

Vol. 4 *Nanoporous Materials: Science and Engineering*
edited by G. Q. Lu & X. S. Zhao

Vol. 3 *Batch Distillation: Design and Operation*
by Iqbal M. Mujtaba

Vol. 2 *Adsorption Analysis: Equilibria and Kinetics*
by Duong D. Do

Vol. 1 *Gas Separation by Adsorption Processes*
by Ralph T. Yang

Series on Chemical Engineering Vol. 6

Metal-Organic Framework Membranes for Molecular Gas Separations

Moises A Carreon
Colorado School of Mines, USA

Surendar R Venna

World Scientific

NEW JERSEY • LONDON • SINGAPORE • BEIJING • SHANGHAI • HONG KONG • TAIPEI • CHENNAI • TOKYO

Published by

World Scientific Publishing Europe Ltd.

57 Shelton Street, Covent Garden, London WC2H 9HE

Head office: 5 Toh Tuck Link, Singapore 596224

USA office: 27 Warren Street, Suite 401-402, Hackensack, NJ 07601

British Library Cataloguing-in-Publication Data
A catalogue record for this book is available from the British Library.

Series on Chemical Engineering — Vol. 6
METAL-ORGANIC FRAMEWORK MEMBRANES FOR MOLECULAR GAS SEPARATIONS

ISBN 978-1-78634-672-8 (hardcover)
ISBN 978-1-78634-673-5 (ebook for institutions)
ISBN 978-1-78634-674-2 (ebook for individuals)

For any available supplementary material, please visit
https://www.worldscientific.com/worldscibooks/10.1142/Q0200#t=suppl

Typeset by Stallion Press
Email: enquiries@stallionpress.com

Preface

Non-thermally driven membrane technology could play a pivotal role in making industrially relevant gas separations less energy intensive and therefore economically feasible. Metal-organic frameworks (MOFs) are microporous crystalline materials consisting on transition metal ions linked together by organic ligands. When prepared in membrane form, MOFs are highly appealing for separating molecular gas mixtures. During the last decade, considerable progress has been made on the development of continuous MOF membranes. In this book, first we present the structural, and compositional properties of MOFs, and justify their use as membranes for gas separation applications. Then, we introduce the parameters and metrics employed for gas separation evaluation.

In the second part of the book, we review representative examples of pure MOF membranes, and MOF mixed matrix membranes that have demonstrated to effectively separate gas mixtures relevant to natural gas purification, carbon dioxide capture, nuclear reprocessing treatment, and hydrocarbon separations among others. We discuss in detail the adopted separation mechanisms over MOF membranes for different gas mixtures. Strategies to synthesize continuous and reproducible membranes are presented.

Finally, we summarize the key aspects that require attention from the scientific community to envision these membranes as an economic viable technology to effectively separate gas mixtures.

With this book, we hope to motivate the scientific community to actively pursue the development of these exciting and promising microporous crystalline membranes for challenging chemical separations.

We are very grateful to Professor Jerry Y.S. Lin (Arizona State University) for all his useful advice and guidance during the process of writing this book.

Moises A. Carreon
Chemical and Biological Engineering Department
Colorado School of Mines, Golden Colorado, USA
mcarreon@mines.edu

Surendar R. Venna
Research Scientist, USA
srvenna@dow.com

Contents

Chapter 1

Metal-Organic Frameworks History and Structural Features

1.1. Introduction

Porous materials are of scientific and technological interest because of their special properties such as high specific surface area, pore volume and uniform pore size and tunable surface chemistry. These porous materials have shown great potential in several industrial applications including molecular separations, gas storage, sensors, heterogeneous catalysis, and drug delivery. Researchers in the fields of chemistry, chemical engineering and materials science are still working to find novel porous materials to further enhance the properties and applications. Zeolites, crystalline aluminosilicates composed of silica, aluminum, sodium, etc, are the prime examples of the porous materials which have been widely used applications in gas separation, ion exchange and catalysis.[1] Zeolites have different pore sizes, shapes and different pore chemistry. The pore size of most zeolites is $<2\,nm$ and depending on the adsorption and/or diffusion kinetics of the gas molecules in the pores, they separate different gas molecules and catalyze the organic molecules depending on their pore chemistry.[2] Although zeolites found great applications in industry, the number of natural or synthesized zeolite are very limited.[3] Also, the chemical functionalization and structure control of this class of porous materials has been challenging. There was a strong need for novel tunable porous structures in the early 1990s. Metal-organic frameworks came to limelight to fulfill the need for pore chemistry tunability.

Metal-organic frameworks (MOFs) are a new class of porous materials comprised of ordered networks consisting of metal ions linked together by organic ligands.[4] Metal-organic frameworks also have similar porous structures compared to zeolites, but can be synthesized boundless variety

Figure 1.1. Omar Yaghi, Professor of Chemistry at UC Berkeley, a pioneer in MOF synthesis. (Reprinted with permission from Professor Omar Yaghi.)

of structures with different surface chemistries and pore structures.[5] MOF kind of structures were reported starting from 1960s by Tomic and others.[6] Since the early 1990s, MOF research has been increased greatly especially with the rediscovery of MOF based porous materials by Yaghi and his research group. Omar Yaghi (Figure 1.1) is a well-known pioneer and leader in developing the chemistry of metal-organic frameworks.[7] He was recently ranked No. 2 among the world's top 100 chemists of the past decades (2000–2010), based on the impact of his published research mostly in the field of MOFs and other porous materials. Their research group is noted for his contribution in introducing metal-oxide clusters as anchors for joining organic linkers into robust crystalline open frameworks with permanent porosity.

This book covers advances in the MOF membrane field over the last two decades, focusing on different gas separation applications. We also discuss several conventional and novel strategies developed by several research groups for the continuous defect-free MOF membrane and film fabrication. The advantages of using MOFs in mixed matrix membranes and improvements in gas separation performances with the MOF-based mixed matrix membranes are also discussed.

1.2. Nomenclature and analogues of MOFs

Metal-organic frameworks really came to light with the discovery of MOF-5 by Professor Yaghi.[8] This MOF-5 broke the historic world record of porosity at that time by showing very high surface area of 6,500 m^2/g. This discovery of MOF-5 is recognized as the most prominent example of a metal-organic framework by many researchers. A major advance in the chemistry of MOFs came with the synthesis of MOF-5. Several properties such as X-ray single crystal structure determination (as shown in Figure 1.2), and gas sorption properties were reported for the first time for a MOF and is the start of the golden age in the field of MOF development. For a solid to be labelled a metal-organic framework, as defined by Yaghi and his coworkers, it should display the inherent attributes that this term implies: strong bonding providing robustness, linking units that are available for modification by organic synthesis, and a geometrically well-defined structure.[4]

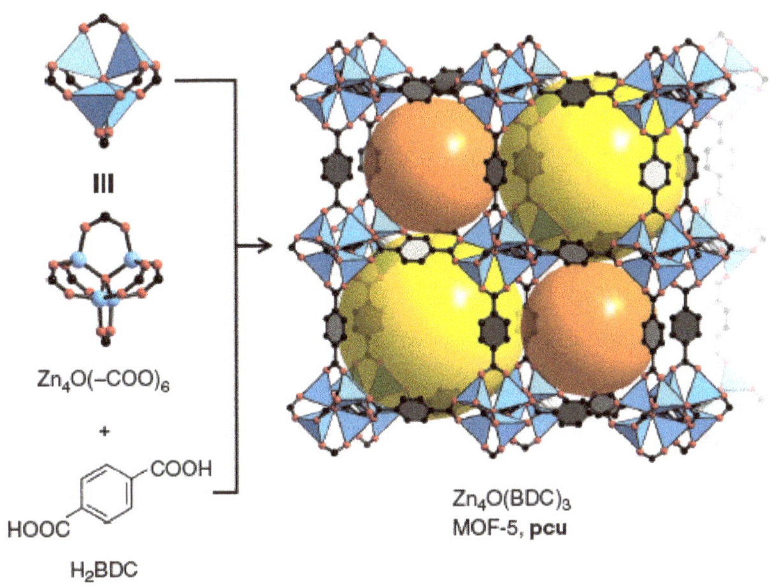

Figure 1.2. Single-crystal X-ray structure of MOF-5 constructed from $Zn_4O(–COO)_6$ SBUs, connected through BDC. The overall structure is composed of a large cavity (15.1 Å, yellow sphere) together with a smaller cavity (11.0 Å in diameter, orange sphere) in an alternating fashion. Hydrogen atoms are omitted for clarity. Color code: black, C; red, O; light blue polyhedra, Zn. (Reprinted from Ref. 9 with permission from the American Chemical Society.)

Some of the MOFs were named based on their chemical structures. MOF-2 (reported in 1998) with formula of Zn(BDC)(H_2O) (BDC = 1,4-benzenedicarboxylate) was one of the first MOFs which exhibited permanent porosity and showed Type-1 adsorption isotherm.[9] This MOF showed Langmuir surface areas between 270 and 310 m^2g^{-1}, as well as micropore volumes between 0.094 and 0.086 cm^3g^{-1}. Such frameworks are useful for making our world sustainable with great short and long-term benefits. Formation of permanent porosity in the MOFs opens up the opportunity for new chemical separations, ion exchange, sensing and catalysis applications.

The researchers gave several names to these kind of porous materials over the last 3 decades because of the lack of a generally accepted definition of this new type of hybrid material as nicely summarized in the review by Li *et al.*[10] Some of them are reported here; porous coordination polymer (PCP) seems to have been the most widely adopted, followed by porous coordination network (PCN). Others include MCP (microporous coordination polymer), ZMOF (zeolite-like metal-organic framework), ZIF (zeolitic imidazolate framework), MPF (metal peptide framework), MAF (metal azolate frameworks), mesoMOF (mesoporous metal-organic framework), and bio-MOF or MBioF (metalbiomolecule framework). The metal-organic frameworks (MOFs) are numbered in their chronological order of discovery in the early stages and later MOFs were named after the universities in which the materials were prepared and researchers who synthesized the materials. For example, UiO-66 (University of Oslo), MIL (Materials Institute Lavoisier), HKUST (Hong Kong University of Science and Technology), ITQMOF (Instituto de Tecnología Química metal-organic framework), SNU (Seoul National University), JUC (Jilin University China), CUK (Cambridge University KRICT), and POST (Pohang University of Science and Technology), etc, are after the university names.

Synthesis of unlimited number of MOFs are possible with different combinations of metal building units and organic linkers. Most striking difference of MOFs compared to other porous materials is that the MOFs are highly tunable. Their structural, magnetic, catalytic, electrical and optical properties are highly tunable to obtain the desired properties. Several thousands of MOF were synthesized for wide variety of applications such as fuel cells, super capacitors, gas storage, ion exchange, gas separation, catalysts, drug delivery and sensing. Wilmer *et al.* developed a computational approach to generate possible MOF structures from a given chemical library of building blocks (based on the structures of known MOFs) and rapidly screened them to find the best candidates for methane storage application.[11] From a library of 102 building blocks, Wilmer *et al.* generated 137,953 hypothetical MOFs and calculated the pore-size distribution, surface area and methane-storage capacity for each of those MOFs. In the last two decades, there were several researchers working on the synthesis of a variety of MOFs, yet only a fraction of the metal-organic linker combinations were studied.

1.3. Progress of MOFs

Metal-organic frameworks (MOFs) are made by linking inorganic and organic units by strong bonds (reticular synthesis). The flexibility with which the constituents, geometry, size, and functionality can be varied has led to more than 20,000 different MOFs being reported and studied within the past decade. The flexibility with which the metal SBUs and organic linkers can be varied has led to thousands of compounds being prepared and studied each year. The number of porous materials such as zeolites, MOFs and other porous materials that are reported in the Cambridge Structural Database (CSD) are increasing at substantial rate in last four decades as shown in Figure 1.3. The number of MOF structure that were reported to the Cambridge Structural Database (CSD) by researchers were also growing exponentially as shown in Figure 1.3.[12] For now there is about 70,000 MOF based materials were reported and trend is going to continue as the development of novel building blocks along with the computational screening of the materials. The number of papers which feature the term 'metal-organic frameworks' has increased dramatically over the last few years as shown in Figure 1.4. Most of the MOF research was focused on the synthesis and development chemistry. Only a limited number of possible applications have been discussed for MOFs so far, and lists of applications are growing at a faster rate.

Polymeric membranes are matured enough and found applications in almost all chemical separation areas because of their low cost, easy processing and very good mechanical properties. But they have their own disadvantages such as short life time, low chemical and thermal stability.

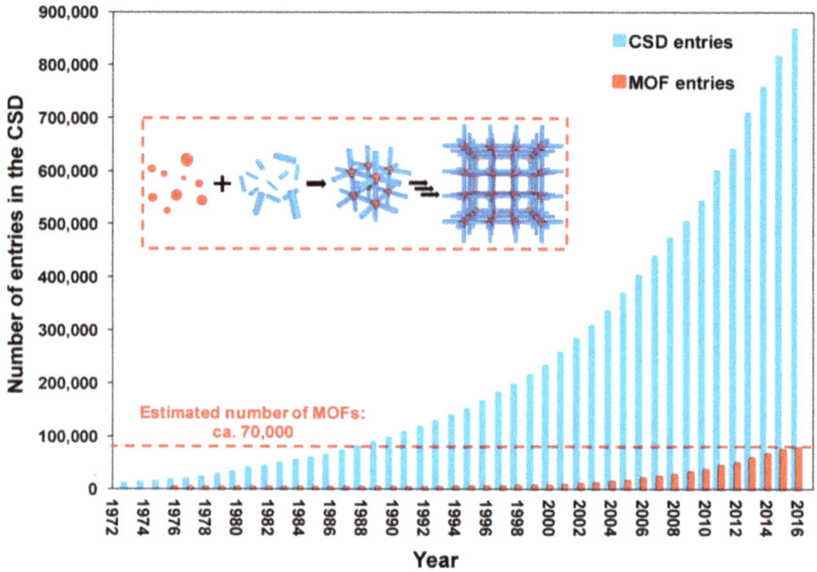

Figure 1.3. Growth of the CSD and MOF entries since 1972. The inset shows the MOF self-assembly process from building blocks: metals (red spheres) and organic ligands (blue struts). (Reprinted from Ref. 12 with permission from the American Chemical Society.)

MOFs in membrane form, have emerged as an appealing type of crystalline microporous materials which combine highly desirable properties, such as uniform micropores, high surface areas, and exceptional thermal and chemical stability, making them ideal candidates for gas separations. The metal-organic frameworks need to be processed into the films to apply them in the membrane industry, sensor industry and others. The research focus in this area of MOF film fabrication is also growing fast as shown in Figure 1.4. Almost all the films made were used as a gas separation membranes except a few of them which were used in other applications like sensors.

MOFs were long thought to be not useful for industrial applications because of their lack of chemical and structural stability and hard to compete with other porous materials interms of cost. Most importantly, cost of MOF synthesis was thought to be very high and it's hard to synthesize in larger batches. But MOFs are one of the fastest growing class of materials in chemistry, having hundreds of research groups around the world

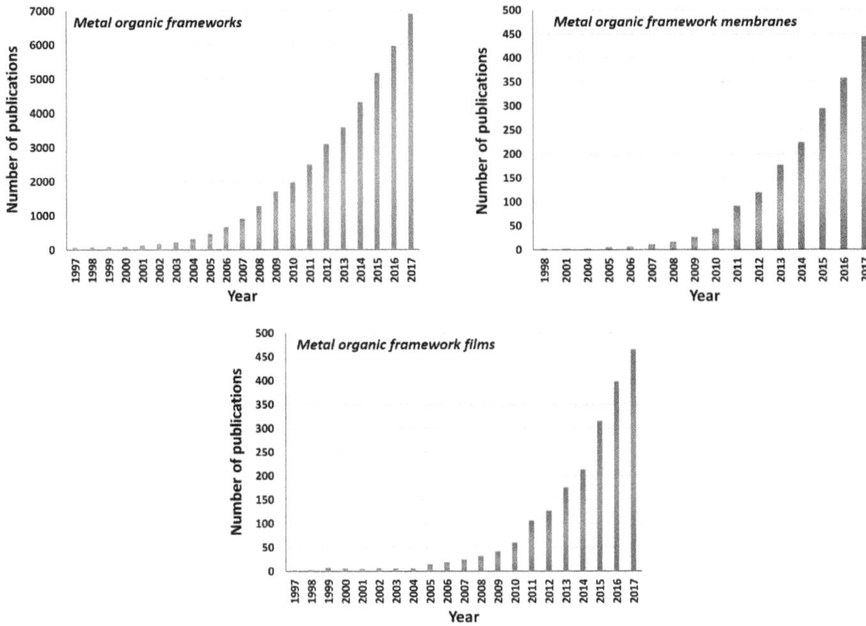

Figure 1.4. Number of publications per year as reported in web of science over last couple of decades; metal-organic frameworks, metal-organic framework membranes and metal-organic framework films.

working aggressively to make robust MOFs in larger quantities. NuMat is one of the MOF-based start-up companies which are producing MOFs in kilogram quantities with less price. Several other companies are also working towards commercialization of these materials such as Mosaic materials, BASF, MOF Technologies, etc, for chemical separations, sensors, and catalysis applications.

1.4. MOF structural generalities

In the past two decades, a new class of crystalline porous materials, MOFs has emerged as one of the most interesting materials. MOF structures consist of metal-containing nodes or cations or clusters linked by organic molecules assembled through coordination bonds to form crystalline 1D, 2D and 3D structures after removal of guest species.[13–15] They can be synthesized inexpensively, relatively easily, in high purity, and in a highly crystalline form.

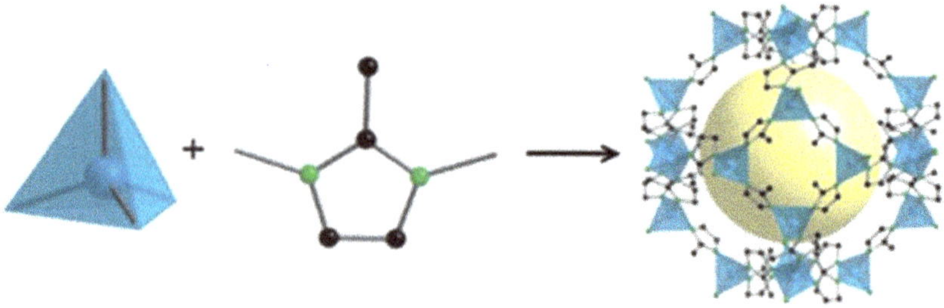

Figure 1.5. A typical MOF structure composed of a metal based cluster and an organic linker. The yellow sphere denotes "open space". (Reprinted from Ref. 16 with permission from Elsevier.)

Different industrial processes needs different properties that are achievable by fine tuning the MOF structures. The combination of organic and inorganic building blocks offers an almost infinite number of variations, enormous flexibility in pore size, shape, and structure, and myriad opportunities for functionalization and grafting which are suitable for a variety of applications. Figure 1.5 shows a typical representation of the components and final structure of a MOF.

The topology of MOFs is related to both the coordination environment favored by the metal ion and the geometry of the organic ligands, to form secondary building units (SBUs) that establish the network symmetry. MOFs can be tailored to specific applications through varying the metals, ligands, and linkers making up the MOF.[17] The presence of both inorganic and organic components enables both the pore size and chemical environment to be tailored to achieve specific properties. This sets MOFs apart from zeolites, which are also crystalline and microporous, but are fully inorganic and thus lack synthetic flexibility. The self-assembly of the metal nodes and organic linkers creates well-defined framework structures with rigid pores which are very robust and do not collapse upon removal of solvent or other "guest" molecules occupying the pores following synthesis resulting in permanent porosity.[18] In some cases of MOFs, the structure collapses upon removal of guest molecules. At the early stages of MOF materials research, there was a doubt that these MOFs ever make into industry for an application due to their structural collapse. But the intensive research by great researchers around the world resulted in chemically strong MOFs which maintained their structures even after the guest molecule removal.[7]

1.5. MOF synthesis and post-synthetic modification

1.5.1. *MOF synthesis*

MOFs are conventionally synthesized by employing a so-called modular synthesis, where crystals are slowly grown from a hot solution by nucleation and growth mechanism to form porous structures.[19] The MOF synthesis chemistry gained lot of interests in the material research community after the report of MOF-5 synthesis by Yaghi and coworkers.[8] MOF-5 structure consists of tetrahedral $[Zn_4O]^{6+}$ clusters bridged by BDC ligands to form a 3D cubic network as shown in Figure 1.6.[5] Several other MOFs, namely isoreticular metal-organic frameworks (IRMOF-1 to IRMOF-16), were synthesized while having similar topology compared to MOF-5.[20] These functionalized derivatives of the MOF-5 were synthesized using other substituted linear dicarboxylate linkers which are of different lengths and different functional groups on the aromatic backbone. These different linkers resulted in different pore sizes, surface areas and different chemical properties. This development gained lots of interests leading several researchers to functionalize and modify these materials to fine tune the properties of MOFs depending on their applications in gas separation, gas storage, sensors and catalysts. This type of study was also extended to various other MOFs, such as UiO-66,[21,22] MIL-53,[23,24] and NOTT-100[25] structure types. This precise control over the synthesis of MOFs to obtain desired properties propelled these materials further to synthesize more advanced MOFs to be usable in several other industrial applications.

Several synthesis methods to obtain metal-organic framework structures have been developed over the last three decades using wide range of parameters such as temperatures, solvent compositions, type of solvent, reagent ratios, reagent concentrations, and reaction times.[5] All of these parameters are very important to obtain the desired phases, structures and properties. Some of the techniques used for the synthesis of MOFs are solvothermal synthesis,[26] microwave heating,[27] sonication assisted synthesis,[28] mechanochemical procedures,[29] solid state synthesis,[30] spray drying[31]

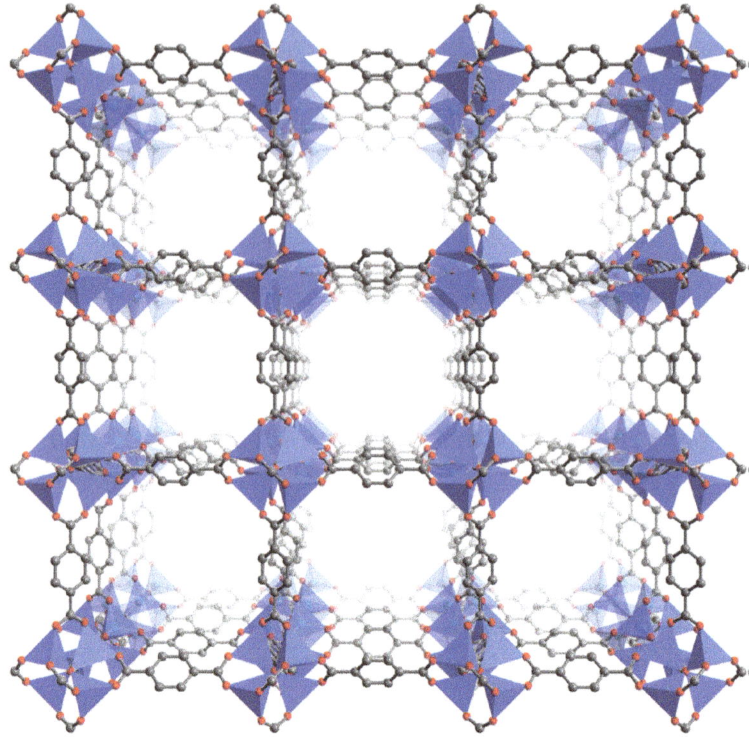

Figure 1.6. A portion of the crystal structure of $Zn_4O(BDC)_3$ (MOF-5). Blue tetrahedra represent ZnO_4 units, while gray and red spheres represent C and O atoms, respectively; H atoms are omitted for clarity. (Reproduced from Ref. 5 with permission from the Springer Nature.)

and electrosynthetic deposition.[32] The timeline for the development of these methods is shown in Figure 1.7. In all of these cases, the solvent occupies the void space within the pores which should be removed carefully using heating or pulling the solvent out using vacuum or by exchanging with other solvents without destroying the structure of the MOFs. As mentioned earlier, the removal of guest molecules results in the open pore structure accessible for other chemical species and is ready to apply for gas separation, catalysis and several other applications.

Over the years, researchers have developed several variety of metal salts and organic linkers to build MOF structures. Deria *et al.* greatly summarized the use of different metal nodes and organic ligands to synthesize various

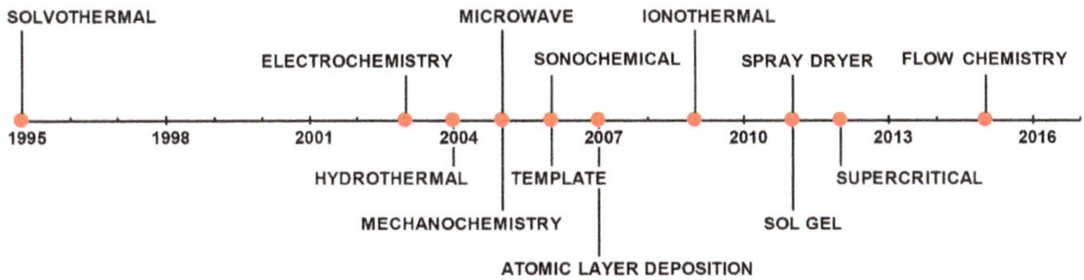

Figure 1.7. Timeline of the most common synthetic approaches patented for the synthesis of MOFs. (Reproduced from Ref. 33 with permission from the Royal Society of Chemistry.)

MOFs as shown in Figure 1.8.[34] HKUST-1 and MIL-based MOFs were prepared using a variety of metal ions (M = Cr, Fe, Zn, Mo) highlighting that the metal is an additional variable in tuning the properties.[5] For MOF's, typical bridging ligands are di- and tricarboxylic acids. These ligands typically have rigid backbones. Some of the examples are benzene-1,4-dicarboxylic acid (BDC), biphenyl-4,4′-dicarboxylic acid (BPDC), and the tricarboxylic acid trimesic acid. Several other organic linkers were used for the synthesis of MOFs as shown in Figure 1.9, which was summarized by Farha *et al.*[35] Till now, only a small portion of possible MOFs were synthesized by experimental researchers. They exist as infinite MOFs comprising of metal nodes and organic linkers connected by coordination bonds of moderate strength as reported by Wilmer *et al.*[11,35] The authors generated the hypothetical MOFs by recombining building blocks derived from crystallographic data of already synthesized MOFs. This approach was compared with snapping Tinkertoys or Lego bricks together.

Even though, the research on MOF synthesis has advanced greatly, still several challenges exists in designing these materials. Slight changes in the reaction parameters employed, such as solvent type, pH, concentration of precursors, presence of a cosolvent, temperature, and reaction time, can change the structural and chemical properties of the obtained MOFs and in some cases completely different MOFs might form. For example, several hundred micron size of a MOF forms using a synthesis procedure, but the synthesis needs to be completely modified to achieve the nanosized crystals of that MOF. Despite the apparent simplicity of the modular synthesis, one of the greatest challenges in preparing new MOFs lies in optimizing the reaction conditions. In order to obtain the desired properties of the reported

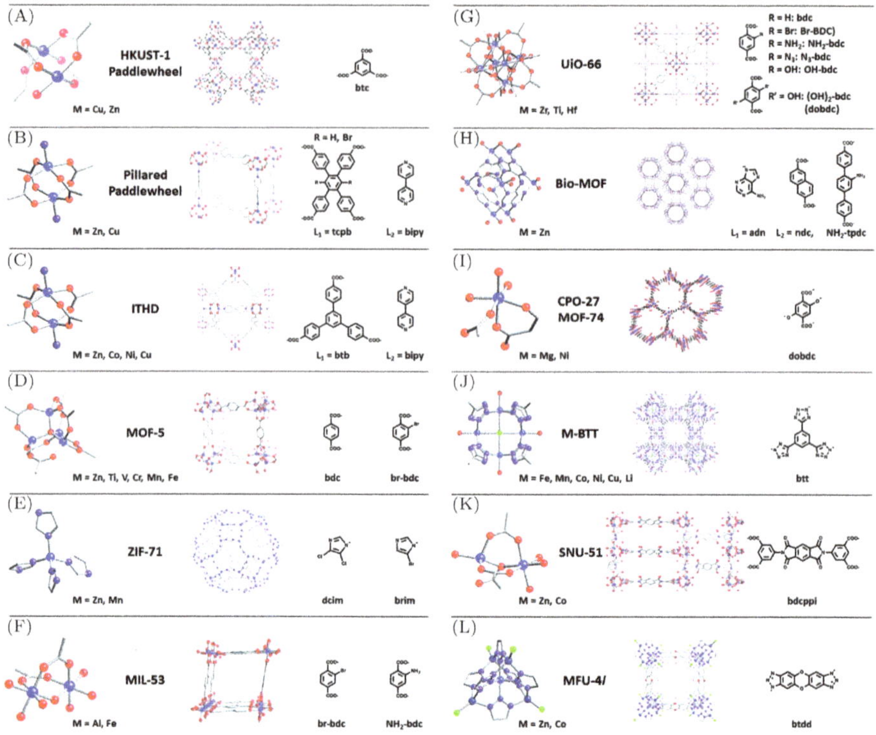

Figure 1.8. Lattice structures (middle) and corresponding SBUs (metal nodes (left), and organic linkers (right)) of some of the MOFs discussed in this review. (Atom definition: blue — metal, red — oxygen, purple — nitrogen, grey — carbon, green — chlorine.) (Reproduced from Ref. 34 with permission from the Royal Society of Chemistry.)

MOF, large number of trial reactions with systematic variation of reaction parameters are needed.[5] In some cases, synthesis of MOFs also results in unwanted side products which compromises the quality of MOFs. Methods such as drygel synthesis, mechanochemical synthesis (which are also called as green methods) were developed to reduce or avoid the formation of undesired products.[36–38]

As the MOF research focus on the synthesis of MOFs with the desired properties towards the targeted applications, MOFs show potential in several industrial applications such as gas storage, gas separation, heterogeneous catalysis, sensors and drug delivery. MOFs show potential in several industrial applications such as gas storage, gas separation, heterogeneous

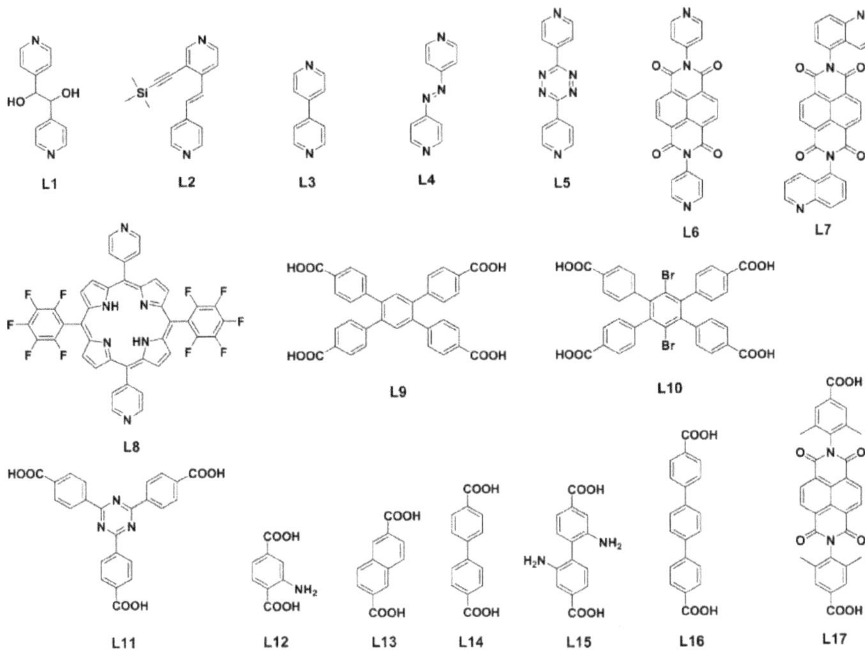

Figure 1.9. Structures of various organic struts employed in the synthesis of MOFs presented in this Account: L1) meso-α,β di(4-pyridyl)glucol; L2) 3-[(trimethylsilyl)ethynyl]-4-[2-(4-pyridinyl) ethenyl]pyridine; L3) 4,4′-dipyridyl; L4) 4,4′-azo-dipyridine; L5) 1,2,4, 5-tetrazine, 3,6-di-(4-pyridinyl); L6) N,N′-di-(4-pyridyl)-1,4,5,8-naphthalenetetracarboxydiimide; L7) N,N′-di-(5-aminoquinoline)-1,4,5,8-naphthalenetetracarboxydiimide; L8) (5,15-dipyridyl-10,20-bis (penta fluorophenyl))porphyrin; L9) 1,2,4,5-tetrakis(4-carboxyphenyl)benzene; L10) 1,4-dibromo-2,3,5,6-tetrakis(4-carboxyphenyl)benzene; L11) 4,4′,4″-s-triazine-2,4,6-triyltribenzoate; L12) 2-aminoterephthalic acid; L13) 2,6-naphthalenedicarboxylate; L14) 4,4′-biphenyldicarboxylic acid; L15) 2,2′-diaminobiphenyl-4, 4′-dicarboxylic acid; L16) 1,4-di(4-carboxyphenyl) benzene; L17) N,N′-di-(3,5-dimethylcarboxyphenyl)-1,4,5,8-naphthalenetetra carboxydiimide. (Reprinted from Ref. 35 with permission from the American Chemical Society.)

catalysis, sensors and drug delivery. For applications such as carbon capture from power plant flue gas and natural gas or hydrogen gas storage, several tons of MOF is needed for these applications. Despite their tremendous potential, near-term prospects for commercial applications remain quite limited because of the lack of technologies and processes for synthesizing these materials in quantities required for industrial applications and at a low cost. But most of the synthesis techniques reported until recently are small scale batch processes. Facile synthesis of MOFs is very important to lower the cost and also to achieve fundamental understanding and viable applications. Also, continuous production of MOF is important to reduce the cost of MOF production instead of batch processing of MOF. Using this continuous process, MOF can be produced with uniform quality and continuously. Compared to the MOF synthesis techniques, solvothermal synthesis is the most commonly studied and used synthetic route. This allows the system to be operated without the need for specialist equipment and the relatively fast growth of crystals with high levels of crystallinity, phase purity and surface areas.[39] Many MOF synthesis require expensive ligands or the use of costly and non-reusable solvents and is adding the cost of MOF production at industrial scale for real world applications.

The issues with large scale production of MOFs were nicely put together by Rubio-Martinez *et al.* in a review published in 2017.[33] Here is the summary of their analysis. The cost, toxicity, corrosive nature and flammability of the organic solvent that were needed for this large scale production of MOF is a big problem. Another major problem is the production of large scale organic ligands needed for MOF synthesis. Other issues for the large scale production of MOFs include anion build up, particle size control, product uniformity and activation of MOFs. The MOF synthesis should be versatile, avoid harsh temperature conditions, continuous process and high space-time-yield parameter.[33]

Recently, several start-up companies like NuMat Technologies, MOF-WORX, MOFapps, novoMOF, Mosaic material and MOF technologies are in the process of producing large quantities of MOFs on the order of kilogram scale. BASF is one of the first companies that focused on the production of MOFs at industrial scale for natural gas storage. In 2016, NuMat Technologies announced the launch of their first product line, ION-X (Figure 1.10) to store the toxic gases such as arsenic, phosphine in the gas cylinder.[40] Similarly, MOF technologies also commercialized their MOFs which are produced by mechanochemical synthesis which uses much less organic solvents while producing the high yields. Mosaic Materials is looking to use MOFs to separate CO_2 gas mixtures from power plant flue gas. The list of MOFs that were developed by different companies is presented in Table 1.1 (Taken from Ref. 33).

Figure 1.10. ION-X gas storage tank to store toxic gas more safely than conventional methods storing. Reprinted with permission from NuMat Technologies Inc. © All rights reserved by NuMat Technologies Inc. ION-X is a registered trademark of NuMat Technologies Inc.

Table 1.1. Sales catalogue of MOFs available by each manufacturer. (Reproduced from Ref. 33 with permission from the Royal Society of Chemistry.)

MOF	Manufacturer
Al(OH) fumarate	MOF Apps
	MOF Technologies
CAU-TO	ProfMOF
Cu-BTC	BASF
	MOF Apps
	MOF Technologies
Fe-BTC	BASF
Magnesium formate	BASF
	MOF Technologies
Mg-MOF-74	MOF Technologies
MIL-100	KRICT
	MOF Apps
MIL-101-NH2	MOF Apps
MIL-53	BASF
MIL-68	MOF Apps
MOF-177	BASF
PGN-250(Fe)	Framergy
UiO-66 series	Inven2
	MOF Apps
	ProfMOF
ZIF-67	MOF Apps
	MOF Technologies
ZIF-8	BASF
	MOF Apps
	MOF Technologies
	STREM Chemicals inc.
Zn-SIFSIX-Pyrazine	MOF Technologies

1.5.2. *Post-synthetic modification (PSM) of MOFs*

MOFs have been synthesized and designed to obtain the desired chemical and physical properties. But some of the MOFs need to be post-modified to obtain the desired properties to fine tune them for a specific application. In the case of other porous materials such as zeolites and activated carbons, this post-synthetic modification is not easy because of the lack of organic linkers. The synthesis of functionalized MOFs using conventional methods is not easy because of high reactivity of the functional groups which reacts to metal centers even before the organic linkers resulting in several undesired and nonporous solids. Post-synthetic modification (PSM) eliminates the need to develop and optimize the precise reaction conditions to form the functionalized MOF directly in a one-pot synthesis.[5] Post-synthetic functionalization allows facile optimization of several properties of MOFs such as pore size, surface area, surface chemistry, chemical and thermal stability. MOFs surface and inside the pores can also be post-synthetically modified without loss of structural integrity, order, or porosity of the extended structure. Post-synthetic modification of MOFs opens up their applications in a wide variety of areas.

Post-synthetic modification of MOFs was broadly divided into three areas by Seth Cohen as published in Chemical Reviews in 2012 as shown in Figure 1.11.[41] In covalent PSM, organic linker or secondary building units are connected with the desired functional group by a new covalent bond. This method is powerful and versatile for the introduction of variety of functional groups. Dative post-synthetic modification was defined as the use of a reagent that forms a metal-ligand bond with a component of the MOF i.e either SBU or the organic linker. Post-synthetic deprotection is another way of modification by the cleavage of a chemical bond within the MOF framework which opens the new functional groups or open metal centers resulting in a MOF with completely different chemical properties. Other methods include solvent assisted linker exchange (SALE), and transmetalation in which organic linkers or metal nodes can be replaced with new linkers or metals to obtain a new framework with the same parent MOF topology.[42] The research work on PSM of MOFs by several research groups across the world showed that MOFs can be functionalized after their synthesis where the composition, structure, functionality, porosity, and metrics of a metal-organic structure can be designed for a specific application.

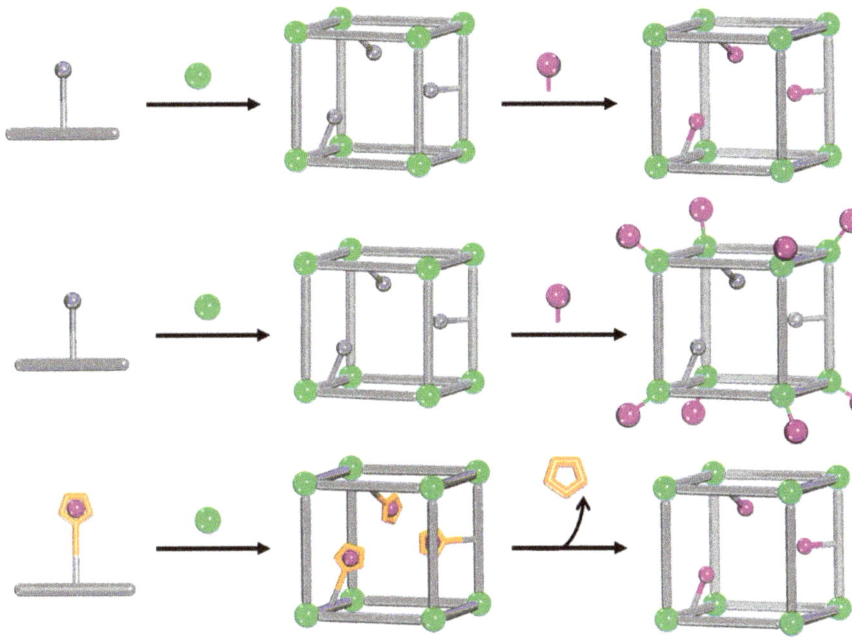

Figure 1.11. Generic schemes for covalent PSM, dative PSM, and Post-synthetic deprotection. (Reprinted from Ref. 41 with permission from the American Chemical Society.)

1.6. Characterization of metal-organic frameworks

As synthesized MOFs and functionalized MOFs needs to be characterized to study their structure, crystallinity, pore size, surface area and chemical properties, these MOF materials needs to be activated carefully to remove any guest molecules that were trapped in the porous structure of MOFs without any destruction of the structure. Vacuum drying, supercritical drying and solvent exchange are the most common methods used to activate the samples.[42] Several characterization techniques were used to study the properties of the MOFs such as powder X-ray diffraction, single crystal X-ray diffraction, scanning electron microscopy, transmission electron microscopy, porosimeter, thermal gravimetric analysis, inductively coupled plasma optical emission spectroscopy (ICP-OES), X-ray photon spectroscopy and Nuclear magnetic resonance. Some of the characterization techniques that were used extensively to study the MOF properties are elaborated here.

1.6.1. *Powder X-ray diffraction*

X-ray diffraction (XRD) is a rapid analytical technique primarily used for phase identification of a crystalline material and can provide information on unit cell dimensions. Powder XRD is a powerful tool to study the network structure and its crystalline properties. The geometry of an X-ray diffractometer is such that the sample rotates in the path of the collimated X-ray beam at an angle θ while the X-ray detector is mounted on an arm to collect the diffracted X-rays and rotates at an angle of 2θ. When X-ray hits the face of a crystal, it diffracts in a pattern characteristic to its structure. In powder X-ray diffraction, the diffraction pattern is obtained from a powder of the material, rather than an individual crystal. Powder diffraction is often easier and more convenient than single crystal diffraction as it does not require individual crystals. A diffraction pattern plots intensity against the angle of the detector, 2θ. For example, powder XRD was used to characterize the formation kinetics of ZIF-8 by measuring the peak intensities as shown in the Figure 1.12.[43] As shown in Figure 2.8, the relative crystallinity of ZIF-8

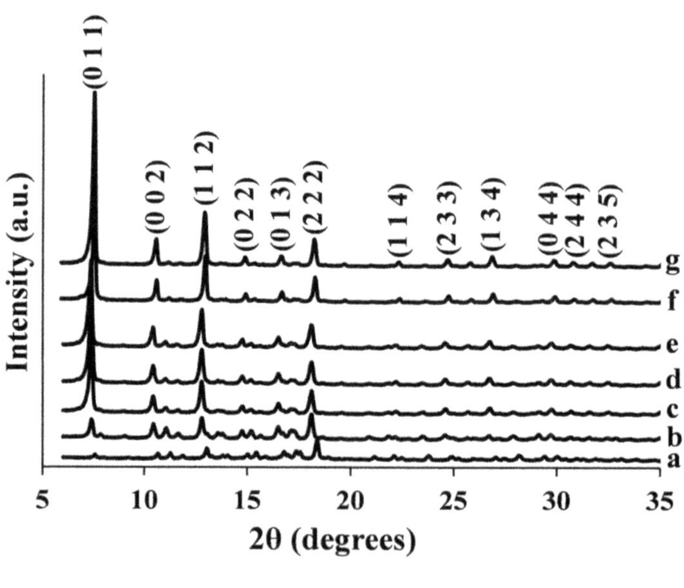

Figure 1.12. XRD patterns illustrating the structural evolution of ZIF-8 as a function of synthesis time: (a) 20 min; (b) 30 min; (c) 40 min; (d) 50 min; (e) 60 min; (f) 12 hrs and (g) 24 hrs. (Reprinted from Ref. 43 with permission from the American Chemical Society.)

increases slowly at initial stages of the reaction. Then a rapid increase in the ZIF-8 formation was observed in the 30 to 40 minutes range. Finally, after 50 min, the *relative crystallinity of ZIF-8* remained practically constant (above 90%), reaching a maximum of 100% at 24 hr.

The structural information of the MOFs such as unit cell size and phase purity were compared with the simulated patterns generated from single crystal X-ray data or through the use of computational modeling.[42] This method also helps to study the MOF structure before and after the post-synthetic modification to confirm its structural integrity. The crystallinity of MOFs also allows precise structural characterization by diffraction methods, thus facilitating their rational design and the formulation of structure–property relationships.

Single-crystal X-ray Diffraction, the most important and straight forward method, provides detailed information about the internal lattice of crystalline substances, including unit cell dimensions, bond-lengths, bond-angles, and atomic thermal displacements and occupancies.[44] Single crystal XRD provides much detailed information about the MOF structures compared to the Powder XRD. The X-rays generated by a cathode ray tube, filtered to produce monochromatic radiation, collimated to concentrate, and directed toward the sample. The interaction of the incident rays with the sample produces constructive interference when conditions satisfy Bragg's Law. Single-crystal refinement, where the data generated from the X-ray analysis is interpreted and refined to obtain the crystal structure of the novel MOFs. Figure 1.13 shows the MOF structural changes developed from single crystal XRD patterns.[45]

1.6.2. *N_2 adsorption/desorption isotherms*

Analyzing the apparent surface area, pore size distribution and pore volume properties of the MOFs is very important in order to applying them in gas separation or storage applications. These properties are measured by the use of N_2 adsorption and desorption isotherms at liquid nitrogen temperature and relative pressures (P/Po) ranging from 0.05–1. The sample needs to dry completely to remove any adsorbed water and other contaminants before measuring the surface area and pore properties.[46] The amount of sample is also important to analyze these properties accurately. As pointed out by Farha and coworkers, the amount of sample (in grams) multiplied by the

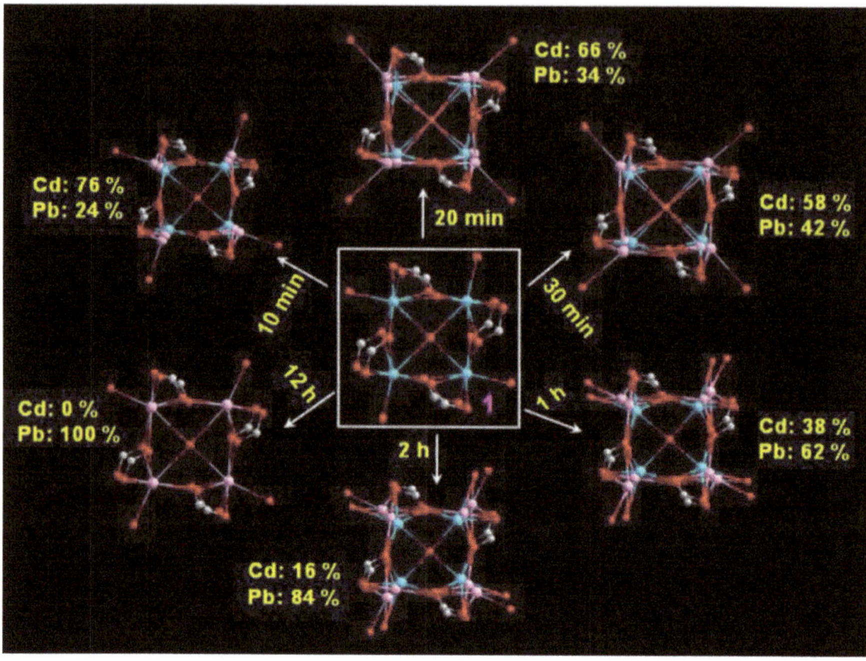

Figure 1.13. The structural changes of the SBU in 1 during the exchange of Cd(II) with Pb(II) as determined by *in situ* single crystal X-ray diffraction. The duration of exchange and the partial occupancy of both Cd(II) and Pb(II) at the SBU are given on the graphics. Some parts of the linker are removed for the clarity. (Reprinted from Ref. 45 with permission from the American Chemical Society.)

specific surface area (in m^2/g) of the sample should be equal to $100\,m^2$ or more to obtain reliable data.[42] A typical isotherm is shown Figure 1.14.[47] The porous properties of NU-1000 based MOFs were also very well matched between the calculated pore volumes and simulated pore property analysis.

The shape of an adsorption/desorption isotherm can also give valuable information about the material. The large uptake of nitrogen at low P/Po indicates filling of the micropores of the MOFs, linear portion of the curve represents multilayer adsorption of N_2, and the concave upward portion of the curve represents filling of larger pores.[48] The surface area of the MOFs is usually measured in the relative pressure range of 0.05 to 0.3. An entire isotherm is needed to calculate the pore size distribution of the MOFs.

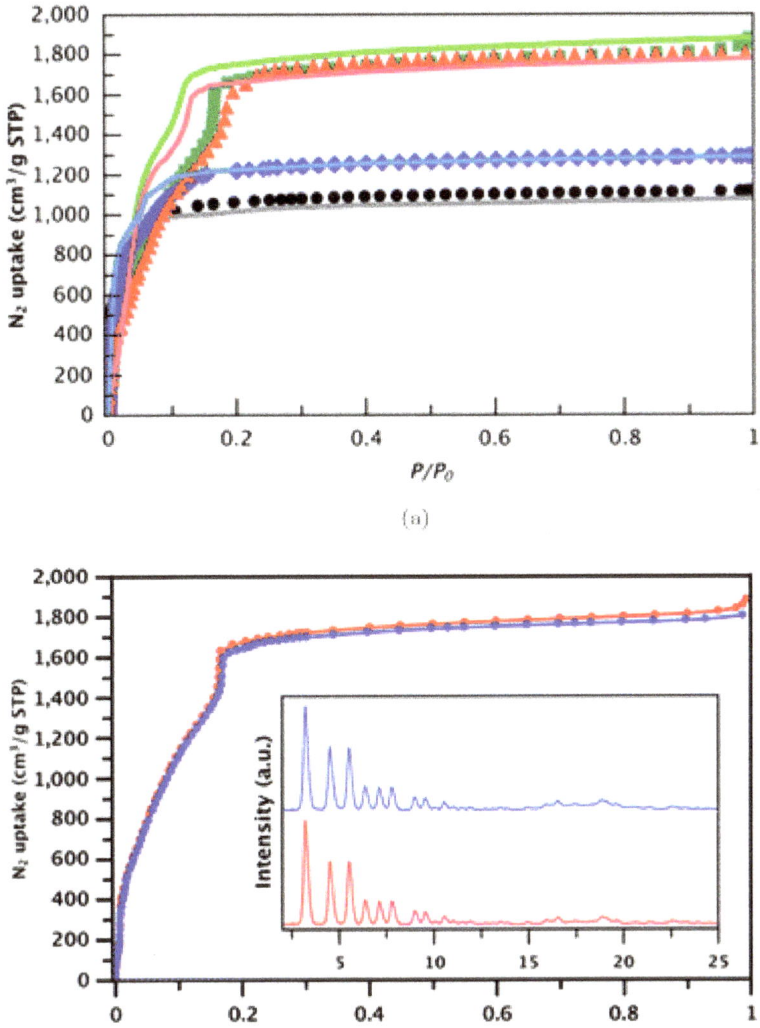

Figure 1.14. (a) Nitrogen isotherms for NU-1101 through NU-1104 (solid lines: simulated isotherm; symbols: experimental isotherm; black: NU-1101, blue: NU-1102, green: NU-1103, red: NU-1104). (b) PXRD patterns (inset) and nitrogen isotherms remained identicalbefore and after water treatment (blue: reactivated sample aftersoaking in water, red: NU-1103right after activation step). (Reprinted from Ref. 47 with permission from the American Chemical Society.)

Most of the researchers used Brunauer–Emmett–Teller (BET) theory[49] for apparent surface area calculations of MOFs since the pore sizes of most MOFs support multilayer gas adsorption. The use of this method for the surface area analysis was reviewed and verified by Walton *et al.* by comparing simulated surface areas of MOFs.[48] The Langmuir theory is valid only for monolayer adsorption and, as a result, tends to overestimate the apparent surface area of most MOFs in some cases by 50% or more.[42] Ideally, five data points, with a minimum of three data points, in the P/P_0 range of 0.05 to 0.30 should be used to successfully determine the surface area using the BET equation (shown below). At relative pressures higher than 0.5, there is the onset of capillary condensation, and at relative pressures that are too low, only monolayer formation occurs.

When the BET equation is plotted, the graph should be of linear with a positive slope.

$$\frac{1}{\left[V\left(\frac{P_0}{P}\right)\right]} = \frac{C-1}{V_m C}\left(\frac{P}{P_0}\right) + \frac{1}{V_m C}$$

where P and P_0 are the equilibrium and the saturation pressure of adsorbates, V is the adsorbed gas quantity, and V_m is the monolayer adsorbed gas quantity. c is the BET constant.

Pore volume and pore size distribution of the MOFs can also be calculated using the sorption isotherms. There are several methods available for the analysis of pore volume and size distributions. Methods based on the Kelvin equation are linked to the pore condensation phenomena, which is useful for mesopore analysis.[50] Methods like Dubinin-Radushkevich approach, and Horvath and Kawazoe (HK) are dedicated to describe micropore filling.[51] Density Functional Theory (DFT) models were developed to analyze these pore properties and is one of the most widely used method especially in the microporous range.[52] X-ray crystal structures were also used to determine the pore properties of the MOFs.

1.6.3. *Termogravimetric analysis (TGA)*

TGA is commonly employed in MOFs research to determine the degradation temperatures, absorbed moisture content, level of inorganic and organic components, and solvent residues. TGA was also used determine the solvent-accessible pore volume. TGA measures the amount and rate of

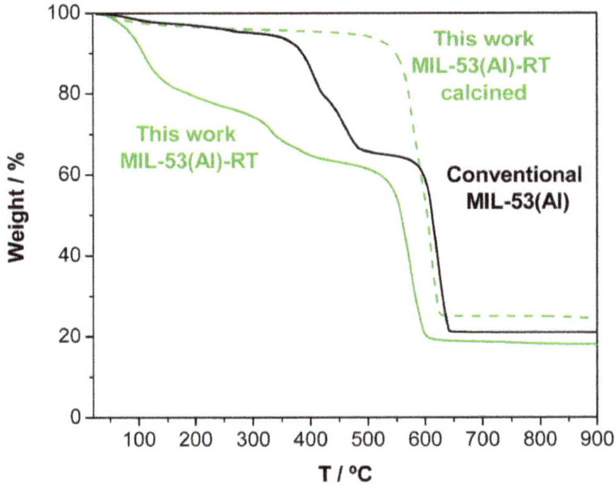

Figure 1.15. TGA curves of the conventional (black line) and RT-prepared (green) MIL-53(Al) materials. The TGA plot of the calcined RT-prepared sample is also shown (green dashed line). (Reproduced from Ref. 53 with permission from the Royal Society of Chemistry.)

change in the mass of a sample as a function of temperature or time in a controlled atmosphere, mostly N_2 gas. In some cases O_2 was also used to study the oxidative properties of the MOFs.

The weight loss occurs due to the changing of chemical or physical properties and minute changes can be detected in the weight of the sample. Typical weight loss curve is shown Figure 1.15.[53] The weight loss curve may require transformation before results may be interpreted. A derivative of the weight loss curve can identify the point where weight loss is most apparent and is very important for the thermal stability analysis of the MOFs.[54] Pore volume of the MOF can also be roughly estimated by collecting the TGA spectrum of a solvated sample.[42] The thermal stability of MOFs is largely determined by its structure and the nature of the chemical bonds. A major concern is that the industrial application of MOFs have been the poor stability of most MOFs. Due to the weak coordination bonds, the thermal stability of MOFs is generally lower than that of zeolites and other porous inorganic solids. The sample must be dried in order to prevent any adsorbed water which can interfere with the thermal stability analysis of the MOFs.

1.6.4. *Scanning electron microscopy (SEM)*

SEM is a useful tool to measure a variety of different MOF properties including particle size, morphology and atomic composition of the samples. SEM uses a focused beam of high-energy electrons to generate a variety of signals at the surface of the sample. Accelerated electrons in an SEM carry significant amounts of kinetic energy, and this energy is dissipated as a variety of signals produced by electron-sample interactions. Out of these signals, secondary electrons are most valuable for showing morphology (as 2D image) and topography on samples and backscattered electrons are most valuable for illustrating contrasts in composition in multiphase samples.

MOF sample are mostly nonconductive and collect the charges which results in the sample damage and image artifacts. This is not a problem with the conductive sample and usually results in very high resolution images. MOF materials needs to be coated with an ultrathin coating of electrically conducting material, such as gold, palladium, carbon, etc, deposited on the sample either by low-vacuum sputter coating or by high-vacuum evaporation to decrease charge buildup from the electron gun. In order to get a very good image, the parameter such as sample voltage, and scan size, stigmation needs to be changed according to the sample. SEM images of the MOFs is presented in Figure 1.16.[55]

Figure 1.16. SEM images of the as-synthesized Co-MOF, Cu-MOF, In-MOF micro-crystals. (Reprinted from Ref. 55 with permission from the American Chemical Society.)

SEM is also capable of performing analyses of selected point locations on the sample; this approach is especially useful in qualitatively or semi-quantitatively determining chemical compositions using Energy Dispersive X-Ray Spectroscopy (EDS). In most cases, the SEM couples with EDS capability. Farha and coworkers pointed out that the coating material choice is very important in obtaining accurate compositions of MOFs.[42] EDS can be used for the preliminary analysis of the chemical composition of MOF since the error percentage is always higher. Other characterization techniques like inductively coupled plasma optical emission spectroscopy (ICP-OES) and X-ray photon spectroscopy can be used to analyze the chemical compositions of the MOFs very accurately.[56] ICP-OES and XPS can also be used to study the impurities present in the MOFs.

Transmission electron microscopy (TEM) is another powerful instrument to study the MOFs to obtain detailed morphology and crystal facets which is not possible with SEM. TEM is especially very useful in the case of nanocrystalline MOFs. TEM is also useful to identify the surface facets and defects in the crystal structure. Even though TEM is very useful, there are only very few researchers who are able to report the TEM analysis of MOFs, mostly due to the instability of these MOFs under TEM electron beam. There is a very nice review which reported the use of TEM to study the crystal structures of MOFs and sample preparation techniques to analyze the MOF under TEM beam without destroying the MOFs.[55] TEM is also useful in studying MOFs that consists of multiple metals or linkers and it can reveal the morphology and size distribution of the embedded nanoparticles and the MOF crystals.

The kinetics of ZIF-8 formation was studied using TEM. Selected area electron diffraction patterns (SAED) as a function of synthesis time were collected and traced the formation of ZIF-8 crystallites with time. *d*-spacing and the cubic lattice parameter of the ZIF-8 particles was also calculated using these patterns.[43] Figure 1.17 shows single crystal SAED patterns from a few randomly selected ZIF-8 crystals. All these patterns were consistently indexed using bcc crystal structure and lattice parameter obtained from these patterns agrees within the experimental error with the one expected for ZIF-8.

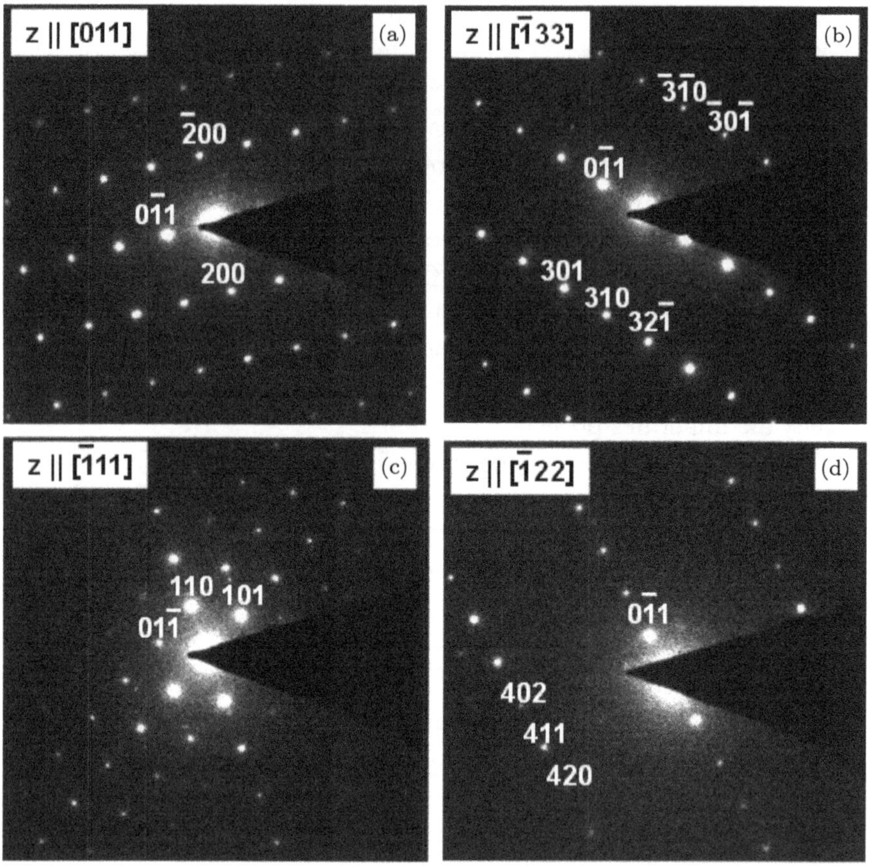

Figure 1.17. SAED patterns of the ZIF-8 crystals synthesized at 12 h taken from four crystals oriented close to: (a) [0 1 1]; (b) [-1 3 3]; (c) [-1 1 1]; and (d) [-1 2 2] zone axis. The indexing of the patterns is consistent with the bcc crystal structure of ZIF-8. (Reprinted from Ref. 43 with permission from the American Chemical Society.)

References

1. Na, K. *et al.* Directing zeolite structures into hierarchically nanoporous architectures. *Science* **333**, 328–332 (2011).
2. Davis, M. E. Ordered porous materials for emerging applications. *Nature* **417**, 813–821 (2002).
3. Wang, Z., Yu, J. & Xu, R. Needs and trends in rational synthesis of zeolitic materials. *Chem. Soc. Rev.* **41**, 1729–1741 (2012).
4. Rowsell, J. L. C. & Yaghi, O. M. Metal-organic frameworks: A new class of porous materials. *Microporous Mesoporous Mater.* **73**, 3–14 (2004).

5. Sumida, K. *et al.* Carbon dioxide capture in metal-organic frameworks. *Chem. Rev.* **112**, 724–781 (2012).
6. Tomic, E. A. Thermal stability of coordination polymers. *J. Appl. Polym. Sci.* **9**, 3745–3752 (1965).
7. Peplow, M. Materials science: The hole story. *Nature* **520**, 148–150 (2015).
8. Hailian Li, Mohamed Eddaoudi, O'Keeffe, M., & Yaghi, O. M. Design and synthesis of an exceptionally stable and highly porous metal-organic framework. *Nature* **402**, 276–279 (1999).
9. Hailian Li, Mohamed Eddaoudi, Thomas L. Groy, & Yaghi, O. M. Establishing Microporosity in Open Metal-Organic Frameworks: Gas Sorption Isotherms for Zn(BDC) (BDC = 1,4-Benzenedicarboxylate). *J. Am. Chem. Soc.* **120**, 33 (1998).
10. Li, J.-R., Sculley, J. & Zhou, H.-C. Metal-organic frameworks for separations. *Chem. Rev.* **112**, 869–932 (2012).
11. Wilmer, C. E. *et al.* Large-scale screening of hypothetical metal-organic frameworks. *Nat. Chem.* **4**, 83–89 (2012).
12. Moghadam, P. Z., Li, A., Wiggin, S. B., Tao, A., Maloney, A. G. P., Wood, P. A., Ward, S. C. & Fairen-Jimenez, D. Development of a Cambridge Structural Database subset: A collection of metal–organic frameworks for past, present, and future. *Chem. Mater.* **29**(7), 2618–2625 (2017).
13. Yaghi, O. M. *et al.* Reticular synthesis and the design of new materials. *Nature* **423**, 705–714 (2003).
14. Férey, G. Hybrid porous solids: past, present, future. *Chem. Soc. Rev.* **37**, 191–214 (2008).
15. MacGillivray, L. Metal Organic Framework: Design and Application, John Wiley & Sons. (2010).
16. Venna, S. R. & Carreon, M. A. Metal organic framework membranes for carbon dioxide separation. *Chem. Eng. Sci.* **124**, 3–19 (2015).
17. Mueller, U. *et al.* Metal-organic frameworks — prospective industrial applications. *J. Mater. Chem.* **16**, 626 (2006).
18. Yaghi, O. M., Li, G. & Li, H. Selective binding and removal of guests in a microporous metal-organic framework. *Nature* **378**, 703–706 (1995).
19. Sharmin, E. & Zafar, F. Introductory Chapter: Metal Organic Frameworks (MOFs). in *Metal-Organic Frameworks* (InTech, 2016). doi:10.5772/64797
20. Eddaoudi, M. *et al.* Systematic design of pore size and functionality in isoreticular MOFs and their application in methane storage. *Science* **295**, 469–472 (2002).
21. Garibay, S. J. & Cohen, S. M. Isoreticular synthesis and modification of frameworks with the UiO-66 topology. *Chem. Commun.* **46**, 7700 (2010).
22. Cavka, J. H. *et al.* A new Zirconium inorganic building brick forming metal organic frameworks with exceptional stability. *J. Am. Chem. Soc.* **130**, 13850–13851 (2008).
23. Biswas, S., Ahnfeldt, T. & Stock, N. New Functionalized Flexible Al-MIL-53-X (X = -Cl, -Br, $-CH_3$, $-NO_2$, $-(OH)_2$) Solids: Syntheses, Characterization, Sorption, and Breathing Behavior. *Inorg. Chem.* **50**, 9518–9526 (2011).
24. Christian Serre, *et al.* Very Large Breathing Effect in the First Nanoporous Chromium(III)-Based Solids: MIL-53 or CrIII(OH)· {O2C–C6H4–CO2} · {HO2C–C6H4–CO2H}x · H2Oy. (2002). doi:10.1021/JA0276974

25. Lin, X. *et al.* High capacity hydrogen adsorption in Cu(II) Tetracarboxylate framework materials: The role of pore size, ligand functionalization, and exposed metal sites. *J. Am. Chem. Soc.* **131**, 2159–2171 (2009).
26. Rosi, N. L., Eddaoudi, M., Kim, J., O'Keeffe, M. & Yaghi, O. M. Infinite secondary building units and forbidden catenation in metal-organic frameworks. *Angew. Chemie* **114**, 294–297 (2002).
27. Ni, Z. & Masel, R. I. Rapid Production of Metal-Organic Frameworks via Microwave-Assisted Solvothermal Synthesis, *J. Am. Chem. Soc.* 128, 38, 12394–12395 (2006).
28. Tehrani, A. A., Safarifard, V., Morsali, A., Bruno, G. & Rudbari, H. A. Ultrasound-assisted synthesis of metal-organic framework nanorods of Zn-HKUST-1 and their templating effects for facile fabrication of zinc oxide nanorods via solid-state transformation. *Inorg. Chem. Commun.* **59**, 41–45 (2015).
29. Klimakow, M., Klobes, P., Thünemann, A. F., Rademann, K. & Emmerling, F. Mechanochemical synthesis of metal-organic frameworks: A fast and facile approach toward quantitative yields and high specific surface areas. *Chem. Mater.* **22**, 5216–5221 (2010).
30. Chen, L., Shen, Y., Bai, J. & Wang, C. Novel symmetrical coralloid Cu 3D superstructures: Solid-state synthesis from a Cu-carboxylate MOF and their in-situ thermal conversion. *J. Solid State Chem.* **182**, 2298–2306 (2009).
31. Garzón-Tovar, L. *et al.* A spray-drying continuous-flow method for simultaneous synthesis and shaping of microspherical high nuclearity MOF beads. *React. Chem. Eng.* **1**, 533–539 (2016).
32. Al-Kutubi, H., Gascon, J., Sudhölter, E. J. R. & Rassaei, L. Electrosynthesis of metal-organic frameworks: Challenges and opportunities. *ChemElectroChem* **2,** 462–474 (2015).
33. Rubio-Martinez, M. *et al.* New synthetic routes towards MOF production at scale. *Chem. Soc. Rev.* **46**, 3453–3480 (2017).
34. Deria, P., Mondloch, J. E., Karagiaridi, O., Bury, W., Hupp, J. T. & Farha, O. K. Beyond post-synthesis modification: Evolution of metal–organic frameworks via building block replacement. *Chem. Soc. Rev.* **43**, 5896 (2014).
35. Farha, O. K. & Hupp, J. T. Rational design, synthesis, purification, and activation of metal-organic framework materials. *Acc. Chem. Res.* **43**, 1166–1175 (2010).
36. Zhan, G. & Zeng, H. C. Alternative synthetic approaches for metal-organic frameworks: transformation from solid matters. *Chem. Commun.* **53**, 72–81 (2017).
37. Das, A. K., Vemuri, R. S., Kutnyakov, I., McGrail, B. P. & Motkuri, R. K. An Efficient synthesis strategy for metal-organic frameworks: Dry-gel synthesis of MOF-74 framework with high yield and improved performance. *Sci. Rep.* **6**, 28050 (2016).
38. Seetharaj, R., Vandana, P. V., Arya, P. & Mathew, S. Dependence of solvents, pH, molar ratio and temperature in tuning metal organic framework architecture. *Arab. J. Chem.* (2016). doi:10.1016/J.ARABJC.2016.01.003
39. McKinstry, C. *et al.* Scalable continuous solvothermal synthesis of metal organic framework (MOF-5) crystals. *Chem. Eng. J.* **285**, 718–725 (2016).
40. Editorial article. Frameworks for commercial success. *Nat. Chem.* **8**, 987 (2016).
41. Cohen, S. M. Postsynthetic Methods for the functionalization of metal-organic frameworks. *Chem. Rev.* **112**, 970–1000 (2012).

42. Howarth, A. J. *et al.* Best practices for the synthesis, activation, and characterization of metal-organic frameworks. *Chem. Mater.* **29**, 26–39 (2017).
43. Venna, S. R., Jasinski, J. B. & Carreon, M. A. Structural evolution of Zeolitic Imidazolate framework-8. *J. Am. Chem. Soc.* **132**, 18030–18033 (2010).
44. Zhang, J.-P., Liao, P.-Q., Zhou, H.-L., Lin, R.-B. & Chen, X.-M. Single-crystal X-ray diffraction studies on structural transformations of porous coordination polymers. *Chem. Soc. Rev.* **43**, 5789–5814 (2014).
45. Das, S., Kim, H. & Kim, K. Metathesis in single crystal: Complete and reversible exchange of metal ions constituting the frameworks of meta–organic frameworks. *J. Am. Chem. Soc.* **131**(11), 3814–3815 (2009).
46. Venna, S. R. & Carreon, M. A. Highly permeable zeolite imidazolate framework-8 membranes for CO_2/CH_4 separation. *J. Am. Chem. Soc.* **132**, 76–78 (2010).
47. Wang, T. C., Bury, W., Gomez-Gualdron, D. A., Vermeulen, N. A., Mondloch, J. E., Deria, P., Zhang, K., Moghadam, P. Z., Sarjeant, A. A., Snurr, R. Q., Stoddart, J. F., Hupp, J. T. & Farha, O. K. *J. Am. Chem. Soc.* **137**, 3585 (2015).
48. Walton, K. S. W. & Snurr, R. Q. Applicability of the BET Method for Determining Surface Areas of Microporous Metal-Organic Frameworks. (2007). doi:10.1021/JA071174K
49. Brunauer, S., Emmett, P. H. & Teller, E. Adsorption of gases in multimolecular layers. *J. Am. Chem. Soc.* **60**, 309–319 (1938).
50. Haynes, J. M. Pore size analysis according to the Kelvin equation. *Matériaux Constr.* **6**, 209–213 (1973).
51. Saha, D., Deng, S. & Yang, Z. Hydrogen adsorption on metal-organic framework (MOF-5) synthesized by DMF approach. *J. Porous Mater.* **16**, 141–149 (2009).
52. Torrisi, A., Bell, R. G. & Mellot-Draznieks, C. Functionalized MOFs for enhanced CO_2 Capture. *Cryst. Growth Des.* **10**, 2839–2841 (2010).
53. Sánchez-Sánchez, M. *et al.* Synthesis of metal-organic frameworks in water at room temperature: salts as linker sources. *Green Chem.* **17**, 1500–1509 (2015).
54. Venna, S. R. *et al.* Fabrication of MMMs with improved gas separation properties using externally-functionalized MOF particles. *J. Mater. Chem. A* **3**, 5014–5022 (2015).
55. Sarawade, P., Tan, H. & Polshettiwar, V. Shape- and morphology-controlled sustainable synthesis of Cu, Co, and in metal-organic frameworks with high CO_2 capture capacity. *ACS Sustain. Chem. Eng.* **1**, 66–74 (2013).
56. Hod, I. *et al.* A porous proton-relaying metal-organic framework material that accelerates electrochemical hydrogen evolution. *Nat. Commun.* **6**, 8304 (2015).

Chapter 2

Adsorption on MOFs

2.1. Introduction

Adsorption is a surface-based process in which molecules, ions or atoms accumulate on the surface of a material usually porous materials. Adsorption is driven generally by the reduction in surface tension between the gas and the solid when the gas molecules adheres to the surface of the solid. The adsorption of a gas on an adsorbent is a spontaneous process, which means that it involves a decrease in free energy and entropy, because the gas adsorbed on the surface of the adsorbent is more ordered than in the gas phase. The gases are called the adsorbate and the solid onto which these gases adsorb are called the adsorbents. It is important to understand the fundamental relation between the adsorbents and gases to enable the rational design and industrialization of adsorbents. Adsorbents can be amorphous or crystalline with different pore sizes and different surface polarity. Adsorption separation mechanisms are highly dependent on the properties of adsorbates and adsorbents used, as well as the associated interactions. Mixture of gases are separated based on the preferential adsorption or accumulation of a gas from a mixture of gases on to the surface of the porous solids.

The common adsorbent for the separation of gas mixtures are porous solids which are characterized by surface area, pore size, pore volume and surface polarity. Different type of porous adsorbents are showed in Figure 2.1.[1] The surface area of the porous materials should be very high to provide the adsorption sites resulting in high adsorption capacity. Microporous materials such as zeolites, activated carbon and metal organic frameworks shows high surface area with small pores whose size is in the range of gas molecule sizes. Different materials provide different pore sizes which can separate different gas molecules by restricting the adsorption of bigger gas molecule than the pore size of the adsorbent.

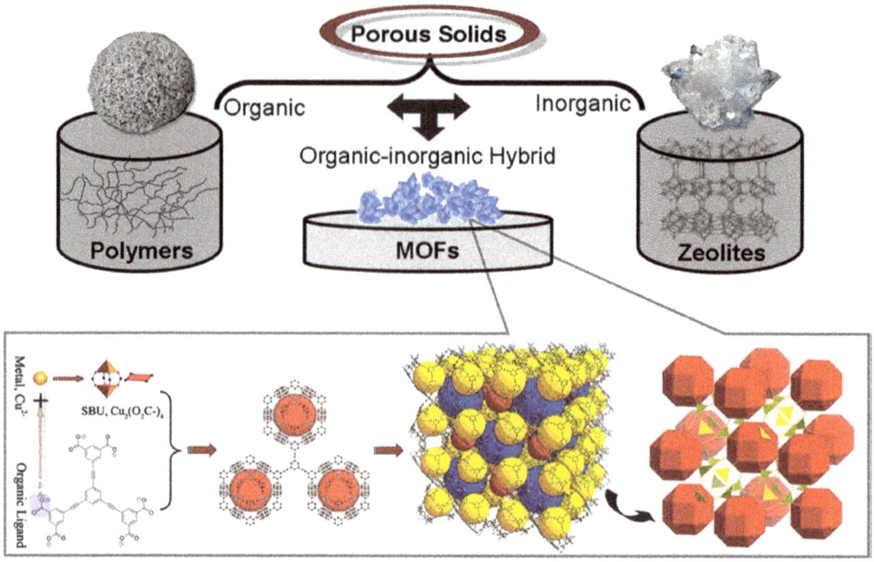

Figure 2.1. Schematic representations of the general classification of porous solids (top, an example is given in each case: polymers for porous organic solids; zeolites for porous inorganic solids; and MOFs for porous organic inorganic hybrid solids) and a typical construction procedure of a MOF (bottom). (Reprinted from Ref. 1 with permission from the American Chemical Society.)

The adsorption process can be classified as physical adsorption and chemical adsorption depending on the strength and nature of the bonds that are formed between adsorbate and adsorbent surface. The physisorption occurs due to the van der Waals force and electrostatic forces (dipole moment and quadrapole moment) between the adsorbent surface and adsorbate. The gas molecules on the surface of the porous materials cover the entire surface of the adsorbent due to the weak Van der walls forces without changing the atomic or molecular orbitals of the gases. Chemisorption involves formation of strong chemical bonds by exchanging or sharing of electrons between gas and solid. Physical adsorption has low heat of sorption compared to the chemical adsorption and need very low energy for the regeneration of the adsorbent.

The potential application of an adsorbent depends on the adsorption equilibria, selectivity and kinetics of the gas adsorption and desorption. Adsorption equilibria depends on the nature of the adsorbate, the nature of the pores in the adsorbent, and the separation process conditions such as temperature and pressures. The selectivity towards a gas depends on the separation process conditions along with chemistry of the adsorbent and adsorbate. Kinetics of the gas adsorption depends on the heat of adsorption and pore size and shape of the adsorbent. Another important parameter that was not widely discussed in the literature is the working capacity of the adsorbent. Working capacity defined as the difference between the amount adsorbed at the process condition and that leaves the adsorbent during purging step. The working capacity of the adsorbent, which depends on the desorption kinetics, should be as high as possible to maximize the separation of gas mixture and to minimize the cost of gas separation.

The adsorption based separation of gas mixtures has several commercial applications in the chemical, petrochemical, environmental, medical and electronic gas industries. Some of the applications include the carbon capture and storage, hydrogen and methane storage, petrochemical separations, air separation, dehydration, impurity removal, pollution control and others across a large number of industrial sectors.[2] Two main process that are used by the industries for gas separation using adsorption are temperature swing adsorption (TSA) and pressure swing adsorption (PSA) processes. PSA has typically been preferred over TSA due to the ease of operation and reliability. In the PSA processes, the gases are selectively adsorbed at higher pressures and desorbed at lower pressures. Most adsorption processes use fixed beds, but some use moving- or fluidized-beds or wheels. As shown in Figure 2.2, the gases are adsorbed on the adsorbents at low

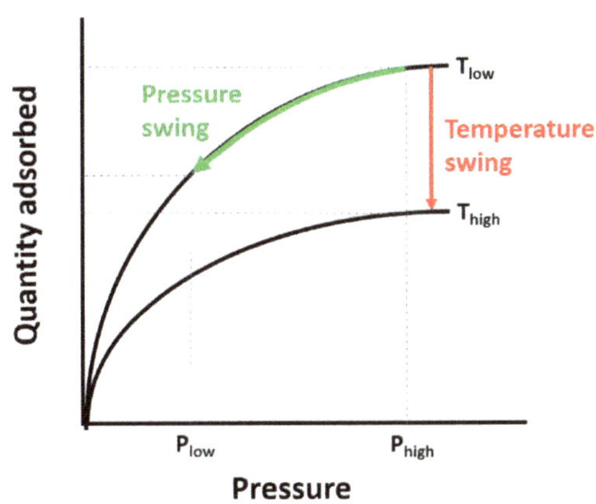

Figure 2.2. Typical adsorption isotherms of PSA and TSA processes.

temperature in a TSA process and the adsorbents are heated to release the adsorbed gases resulting in the rich stream of desired gas. The desorption or regeneration temperature of a gas in TSA process is largely depends on the thermodynamics of adsorption, usually expressed as heat of adsorption or binding energy of adsorption. Heat of adsorption is usually calculated by measuring the adsorption equilibrium at different temperature and calorimetric methodology as clearly discussed by Hu *et al.*[3] The binding energy should be optimum to minimize the temperature difference between the adsorption and desorption process.

A variety of promising types of porous materials such as activated carbon, silica and zeolites have been extensively studied and applied in the industry for adsorption-based separations. The tunability and scope of further development is very limited due to their structural properties. Significant progress has been made recently in developing the novel adsorbents for gas separation to improve the adsorption capacity and increasing the selectivity towards the desired gas which results in reduced cost of separation. Metal-Organic Frameworks (MOFs) have gained lot of interest due to their impressive high surface areas and most importantly, their tunability of the properties such as surface chemistry, pore size and volume. MOF materials can be designed based on their application requirement. They are very easily tunable compared to the other traditional adsorbent materials and can be applied in a wide variety of industrial applications.

2.2. Characterization of adsorbents

The separation performance of an adsorbent mainly depend on the pore size, pore volume, surface area, adsorption capacity and desorption kinetics. Characterization of these properties is very important to analyze their potential for practical applications. Currently, major adsorption separation-based gas separation literature reported their applicability based on the single-component gas sorption isotherms which are important to understand the microscopic sorption sites and properties of gas molecules. It is also very important to analyze these adsorbents using the targeted gas mixtures including the contaminants.

The adsorbents are characterized usually by adsorption isotherms analysis, breakthrough measurement and working capacity calculation. The size of the micropores is very important for gas separation application since pore size can screen the gas molecules depending on the size of the gas molecule and determines the accessibility of gas to the internal pore surface of the adsorbents. The structures of the porous materials such as zeolites and MOFs are crystalline. It is easy to calculate the pore size and its distribution, pore structure and surface area from its crystallographic structures. MOFs are well known materials for pore size tunability accurately and engineered porosity. Measurement of isosteric heat of adsorption is another important property to determine the regeneration process.

The textural properties of the metal organic frameworks are usually characterized by measuring the Nitrogen adsorption-desorption isotherms at 77 K using Brunauer–Emmett–Teller (BET) method as discussed in Chapter 1. It is widely discussed that the N_2 adsorption measurement is not an accurate method to calculate the microporosity because of the accessibility of nitrogen gas in some of the small pore or ultramicropore MOFs and it takes a very long time to get equilibrium with N_2 gas. Consequently, alternative probe molecules such as argon and carbon dioxide were used to measure the microporosity properties more precisely. Both argon and carbon dioxide can enter the small pores of the metal-organic frameworks and achieve the equilibrium very fast compared to nitrogen gas. Nonpolarity of the Argon gas molecules is further advantageous since Argon doesn't interact with the pore surface functional groups and the ions. Moellmer and his collegues discussed the advantages of using argon compared to N_2 to measure the micropore size in their publication.[4] The quadrapole nature of the nitrogen molecule results in the orientation of sorbed molecule on the polar hydroxyl surface group. The effective cross-sectional area of the oriented nitrogen molecules is smaller than the actual surface area of the molecule. Since the argon molecule is monoatomic and nonpolar than the nitrogen, the functional groups on the surface has no effect on the cross-sectional area of the argon and results in accurate structural pore property analysis.

Pore structure and surface analysis is very important to understand the fundamental properties of the material. But, for practical applications, measurement of gas adsorption isotherms on the metal-organic frameworks is very important. The measurement of adsorption isotherms at different temperatures of two different MOFs by Mason *et al.* is a great example to discuss the importance of this measurement for practical applications.[5] Mason *et al.* measured the adsorption isotherms of MOF-177 and Mg_2(dobdc) every

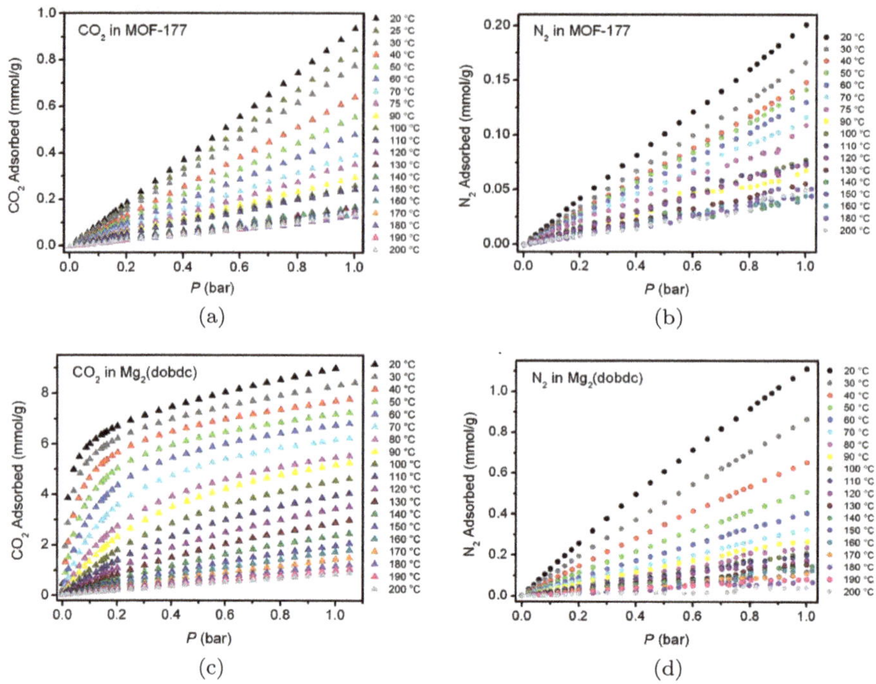

Figure 2.3. Excess CO_2 (triangles) and N_2 (circles) adsorption isotherms for MOF-177 (a and b) and Mg_2(dobdc) (c and d) measured from 20 °C to 200 °C. (Reproduced from Ref. 5 with permission from the Royal Society of Chemistry.)

10 °C from 20 °C to 200 °C as shown in Figure 2.3. The adsorption quantity is decreased with increase in temperature for both gases due to the higher energy of gas molecules at high temperature. This is the usual behavior for almost all adsorbents such as zeolite, activated carbon and metal-organic frameworks. The adsorption capacity for CO_2 is higher compared to N_2 for the adsorbents studied by Mason *et al.* due to higher quadrapole moment and polarizability of CO_2. For practical TSA process, it is important to study the regeneration temperature of the adsorbents. Mason *et al.* also calculated the isosteric heat of adsorption by fitting the adsorption isotherms at different temperature dual site Langmuir model. Isosteric heat of adsorption represents the binding energy of gas molecules on the adsorbent surface. For practical application, heat of adsorption should be moderated to make the separation process less energetic.

Equilibrium selectivity is the key parameter to evaluate separation ability of metal-organic frameworks. The higher the selectivity or separation factor, the higher the purity of the desired gas in the product gas. Adsorption selectivity is the ratio of the amount of adsorbed gases on the surfaces of the MOFs at relative partial pressures of the gases in the feed. The selectivity of the MOF towards a gas can be predicted from the experimental single-component isotherms using the Ideal Adsorbed Solution Theory (IAST).[5] IAST became one of the most used methods because of its simplicity. Adsorption analysis can be done with pure gas isotherms, and then predict any combination of the gases. The main assumption of this theory are all adsorbate molecules in the mixture have equal access to the entire surface area of the adsorbent, the adsorbent is homogeneous and the adsorbed phase is an ideal solution in which intermolecular interactions are equivalent in strength.[6]

It is also very important to analyze these adsorbent by conducting breakthrough analysis (concentration of gas vs time) to provide more guidance towards applicability of these metal-organic framework for adsorptive industrial gas separation applications. The breakthrough curve analysis describes the performance of a adsorption column. The time for breakthrough appearance and the shape of breakthrough curve defines the adsorption-based separation process. Liu *et al.* reported the CO_2/N_2 selective adsorption in the Ni/DOBDC using a fixed bed breakthrough method as shown in Figure 2.4.[7] CO_2/N_2 mixture breakthrough measurement performed using an intelligent gravimetric analyzer (IGA) supplied by Hidden Analytical Ltd., Warrington, UK. The IGA system has a dynamic sorption analyzer (ARBC) that can handle breakthrough adsorption experiment for multiple components.[8] Liu *et al.* conducted breakthrough analysis by processing the Ni/DOBDC MOF into pellet samples and loaded them into the fixed adsorption bed by holding them with quartz wool on both sides. The gas mixture was pumped through the bottom of the bed and continuously observed the breakthrough of the effluent gases using a mass spectrometer. Typical breakthrough curves are shown in Figure 2.5 for MIL-101 (Cr) MOFs measured by Liu *et al.* The analysis of breakthrough curves in the presence of water and other contaminants is also important to understand the effect of them in the final application of these adsorbents.

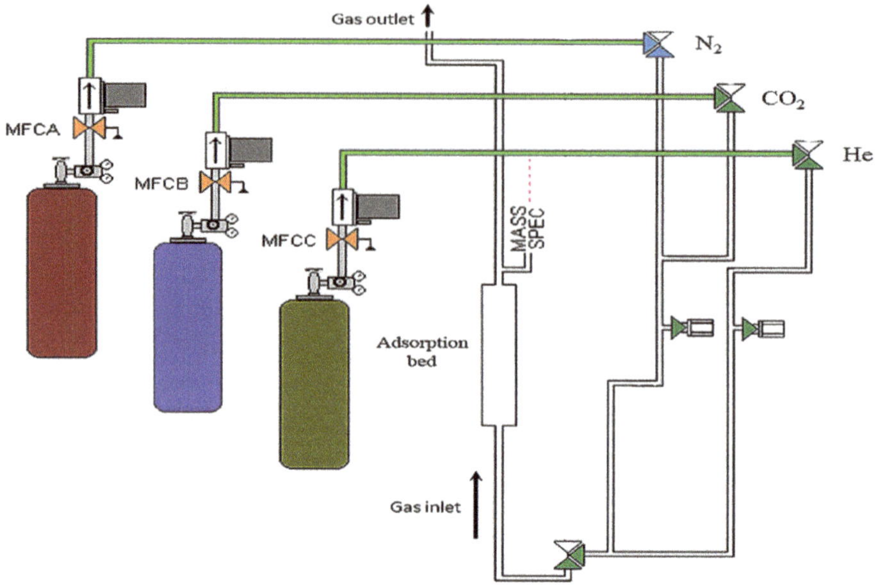

Figure 2.4. Apparatus for the CO_2/N_2 and $CO_2/N_2/H_2O$ mixture breakthrough experiments. (Reprinted from Ref. 7 with permission from the American Chemical Society.)

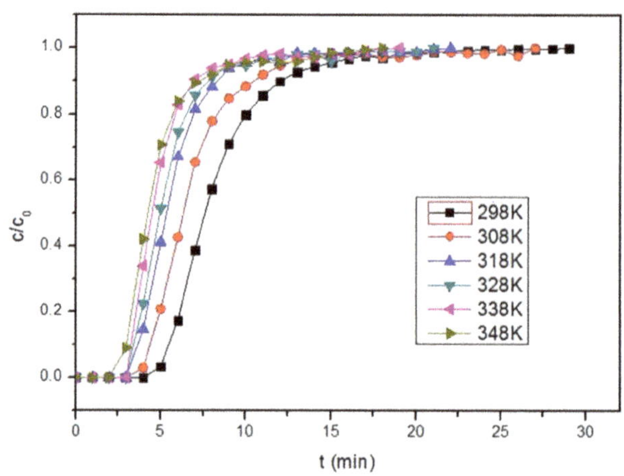

Figure 2.5. Breakthrough curves of 10 vol% CO_2 adsorption on MIL-101(Cr) at different temperatures. (Reproduced from Ref. 9 with permission from the Springer Nature.)

2.3. Designing the MOFs for adsorption-based gas separations

A wide variety of MOFs such as MOF-74, UiO-66, MIL, HKUST, MOF-5, ZIF-8, etc, have been tested in a range of potential applications for gas separations. Most of the earlier metal organic frameworks were not pre-designed as adsorbents or for other applications. MOFs were synthesized by changing the chemistry and analyzed them for different applications — adsorption is one of them. As mentioned in Chapter 1, MOFs can be easily tunable at the molecular level by changing the organic linkers, metal nodes or the synthesis conditions. As the MOF synthesis chemistry developed, many research works are coming out which are specifically designed as the adsorbents for gas separation applications. Although the MOFs can be designed with almost thousands of combinations of linkers and metal nodes, the design should consider its chemical, thermal and mechanical stability for industrial applications.

2.3.1. *Controlling the structural features of MOFs*

Pore size and shape are the first screening point for the applicability of metal-organic frameworks for desired gas separation. The diffusion kinetics entirely depends on the textural properties of the pore and its connectivity in the crystal structure of the MOFs. The textural properties of the MOFs are easily tunable compared to the other porous materials which make them the winner for gas separation applications. The size or length of a ligand is the key in the design of a MOF with desired pore size and shape. The MOF pores can be functionalized and the pore sizes tuned to make them size selective (molecular sieving) for the separation of gases. The pores of MOFs can be designed based on their targeted application and desired gas mixture separation based on the sizes of the gases. Meng *et al.* synthesized three Co-based isostructural MOF-74-III materials were synthesized with three different expanded pore sizes. Extent of fused benzene rings onto side chain of same-length ligands were varied to fine tune the pore sizes to 2.6, 2.4, and 2.2 nm.[10] Gas sorption analysis showed that the alternately arranged fused benzene rings on one side of the ligand served as extra anchoring sites for CO_2 molecules with π–π interactions resulting higher CO_2 adsorption with good CO_2/CH_4 and

CO_2/N_2 selectivity. Shekhah *et al.* showed a material design and engineering approach to control the pore size in combination with suitable energetics of favorable electrostatics from an array of inorganic anions.[11] Novel SIFSIX-3-Cu MOF was synthesized with small pore compared to the parent Zinc-based SIFSIX MOF. SIFSIX-3-Cu resulted in smaller pore of 3.5 Å while Zinc-based SIFSIX MOF has 3.84 Å as shown in Figure 2.6. This enhances the adsorption energetics and subsequently displays higher CO_2 uptake at very low partial pressures. This MOF has the potential industrial

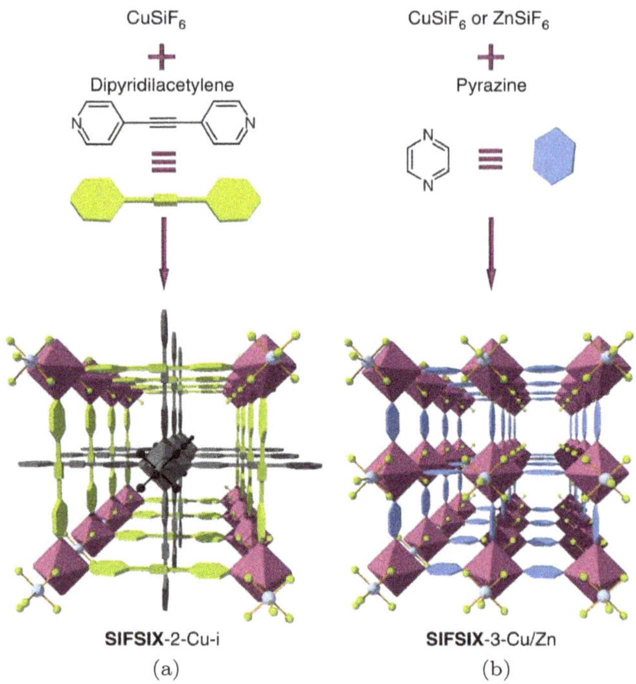

Figure 2.6. Pore size tuning of the channel structures of (a) SIFSIX-2-Cu-i, (b) SIFSIX-3-Zn or/and SIFSIX-3-Cu. (a) SIFSIX-2-Cu-i; pores size 5.15 Å, Brunauer–Emmett–Teller (BET) apparent surface area (N_2 adsorption) 735 m^2/g. (b) SIFSIX-3-Zn; pores size 3.84 Å, BET apparent surface area 250 m^2/g (determined from the CO_2 adsorption isotherm at 298 K); SIFSIX-3-Cu; pores size 3.50 Å (NLDFT), BET and langmuir apparent surface area ca. 300 $m^2 g^{-1}$ (determined from the CO_2 adsorption isotherm at 298 K). Colour code: Dipyridylacetylene (dpa, thick light green polygon), pyrazine (pyz, blue polygon), Zn, Cu (purple polyhedra), Si (light blue spheres), F (light green spheres). All guest molecules are omitted for clarity. Note that the grey net represents the interpenetrated net in SIFSIX-2-Cu-i. (Reproduced from Ref. 11 with permission from the Springer Nature.)

applications such as direct air capture, low CO_2 conc streams from various industries.

In most MOFs, the pore size and pore volumes are fixed due to their rigid crystal structure. But, in some of the metal-organic frameworks (MOFs) structures are very flexible and structural changes occurs in response to external factors such as pressure, temperature, and adsorption of gas or liquid molecules.[12] Crystal structure formation of metal-organic frameworks is composed of weaker coordination bonds and hydrogen bonds which results in the weak bond strength and weak mechanical properties. The physical and chemical properties of these flexible MOFs changes upon exposure to temperature, pressure or gas adsorption. These flexible MOFs shows unusual adsorption isotherms with up to three distinct steps showing the breathing behavior as shown in Figure 2.7.[13] The gas adsorption isotherms in flexible MOFs does not simply follow one of the common six isotherms but is qualitatively distinct from that of rigid microporous materials.[13,14] The adsorption isotherm shows a sudden increase of adsorption at a relatively low pressure by gate opening mechanism and equilibrates at higher pressures. Desorption of gases in these flexible MOFs does not follow the adsorption curve and results in hysteresis.

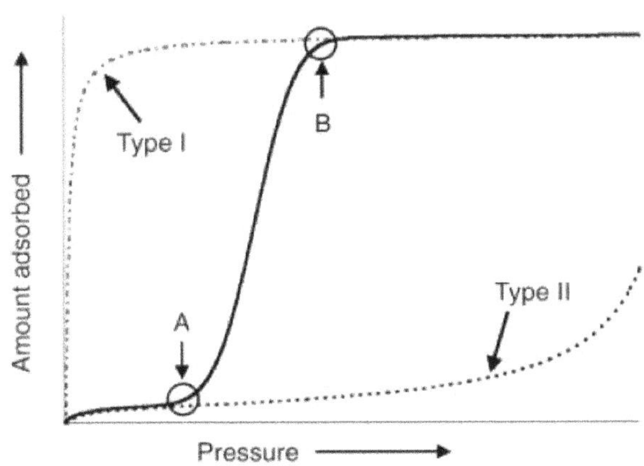

Figure 2.7. Typical adsorption isotherm for flexible PCPs that underwent a structural transformation. Dashed lines represent the type I and type II isotherms. Points A and B indicate the gate-opening and gate-closing pressures which accompany the start and end of structural transformation, respectively. (Reproduced from Ref. 13 with permission from the Wiley.)

MIL (Materials of Institut Lavoisier) series of MOFs are one of the first and well known for its flexibility. MIL-53 (Al) MOFs has attracted a lot of attention due to its flexible nature and its breathing effect during adsorption. MIL-53(Al) is composed of parallel aluminum hydroxide chains connected through 1,4–benzenedicarboxylate (BDC) ligands. This MOF showed the topological freedom to undergo phase transitions with different pore sizes which have a remarkable difference in cell volume.[15,16] Large pore MIL-53 (Al) MOFs is the most stable form at room temperature and in the absence of guest molecules. But at higher temperature and pressure, the large pore MOF transitions into closed pore MOF and it transitions to narrow pore MOF upon adsorption of gases as shown in Figure 2.8.[15] Although there is no industrial application yet, two flexible MOFs were already offered by Sigma-Aldrich under the names Basolites A100 (MIL-53(Al) analogue) and Basolites Z1200 (ZIF-8), correspondingly. The flexible nature of these MOFs makes them excellent candidates for gas separation application. They are applicable for the separation of gases in energetically favorable pressure swing separation processes due to the differences in adsorption of different gases.

The pore size and its connectivity can also be changed by interpenetrating two or more networks of MOF by catenation. Interpenetrating metal-organic frameworks gained a lot of interest because of their fascinating structural features which are very useful for their gas separation applications. The review by Jiang *et al.* provides detailed information about the interpenetrated networks in MOFs and their applications.[17] Interpenetration is the catenation of two or more individual networks with the same or

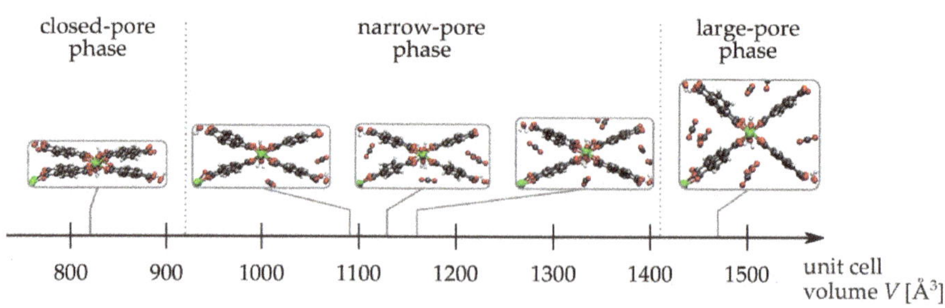

Figure 2.8. The closed-pore (cp), narrow-pore (np), and large-pore (lp) phases of MIL-53(Al) as encountered during CO_2 adsorption. (Reproduced from Ref. 15 with permission from the Wiley.)

different components. The interpenetrating metal-organic frameworks are formed by minimizing the structural energy by filling the void spaces in the porous network. Interpenetration also significantly enhances the stability of the frameworks by minimizing the stresses in the MOF structure and also by the formation of repulsive forces which prevents the collapse of the network.[17] The structural interpenetration may occur only if the pore space of an individual net is sufficiently large to accommodate the organic linkers. The pore sizes can also be controlled by interpenetration networks of the MOFs. Kuang *et al.* synthesized an infinite 3D MOF composed of catenated polyhedral cages, in which the framework is achieved by mechanical interlocking of all of the vertices of the cages as shown in Figure 2.9.[18] Initial MOF research was focused on increasing the surface area for gas storage applications. But the interpenetration serves the exact opposite purpose by decreasing the pore size and volume resulting in lower surface area. But the interaction between the framework and gas molecules might be stronger due to increased electron density resulting in higher gas adsorption at lower gas partial pressures while the non-interpenetrated, large pore MOF shows high gas adsorption at high pressures.

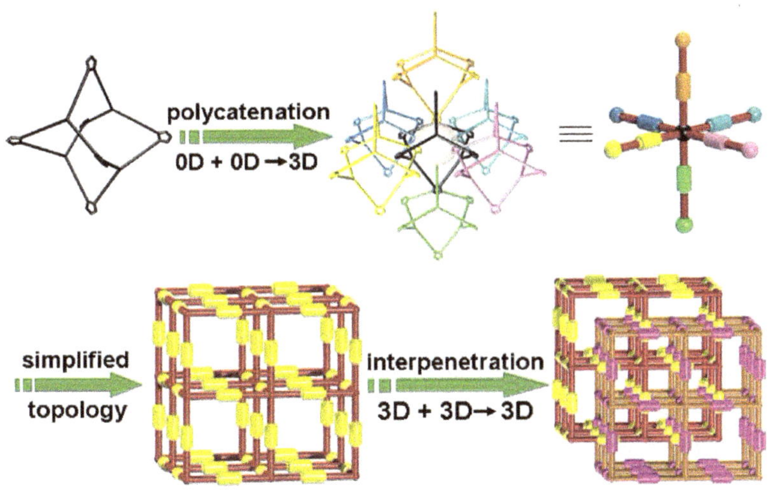

Figure 2.9. Schematic illustration shows "assembly procedure" for the 2-fold interpenetrated framework involving the polycatenation of 0D cages to 3D architectures that are interpenetrated. (Reproduced from Ref. 18 with permission from the Springer Nature.)

2.3.2. *Post-synthetic modification of MOFs for improved adsorption*

Scientists around the world are focusing their efforts on designing the new metal-organic frameworks because of their diverse applicability. Meanwhile, several researchers also focused on improving the gas separation performance of the already synthesized MOF by functionalization. Since most of the MOFs were synthesized using solvothermal methods, it is difficult to incorporate the desired functional groups in its structure without interrupting the crystal structure or modifying the synthesis conditions with care. Post-synthetic modification is an alternative method to attach the desired functional groups to the MOF crystal structure. Cohen and colleagues reviewed the methods that were used to functionalize the MOFs over the last two decades in their review paper.[19,20] The functional groups such as aldehydes, carboxylic acids, nitriles, azides, amines, thiols can be incorporated into the MOF structure. The post-synthetic modification approach is an important tool in developing functionally diverse MOFs with desired properties. Adding functional groups to the MOF crystal structure further enhances the gas adsorption properties. The functionalization MOFs is also very useful for the purification of toxic gas stream by adsorbing the toxic components by chemical interaction.

Nguyen *et al.* functionalized two different MOFs; IRMOF-3 and MIL-53(Al)-NH_2 with long alkyl chains to achieve super hydrophobic surface and to stabilize these materials to moisture.[22] IRMOF-3 (isoreticular metal-organic framework), UMCM-1-NH_2 (University of Michigan crystalline material), and DMOF-1-NH_2 (DABCO metal-organic framework) have been covalently modified with a series of anhydrides or isocyanates to enhance the affinity towards hydrogen adsorption and storage capacity.[23] Alkyl anhydrides modification of DMOF-1-NH_2 induced the flexibility in its structure in the presence of CO_2 and other gases. Kandiah *et al.*[24] and Venna *et al.*[25] first synthesized the amine containing UiO-66-MOF and functionalized with variety of acid anhydrides using amine handle. Marshall and Fogan nicely summarized the functionalization scheme of UiO-66-NH_2 MOF as shown in Figure 2.10.[21] Copper(II) triazole MOF was functionalized with ethylenediamine by exposing the MOF to ethylenediamine contained toluene solutions.[26] Due to the coordination of metal center with the ethylenediamine, the color of the MOF was changed from red to blue. Ethylenediamine functionalization of this MOF resulted in higher CO_2/N_2 adsorption selectivity.

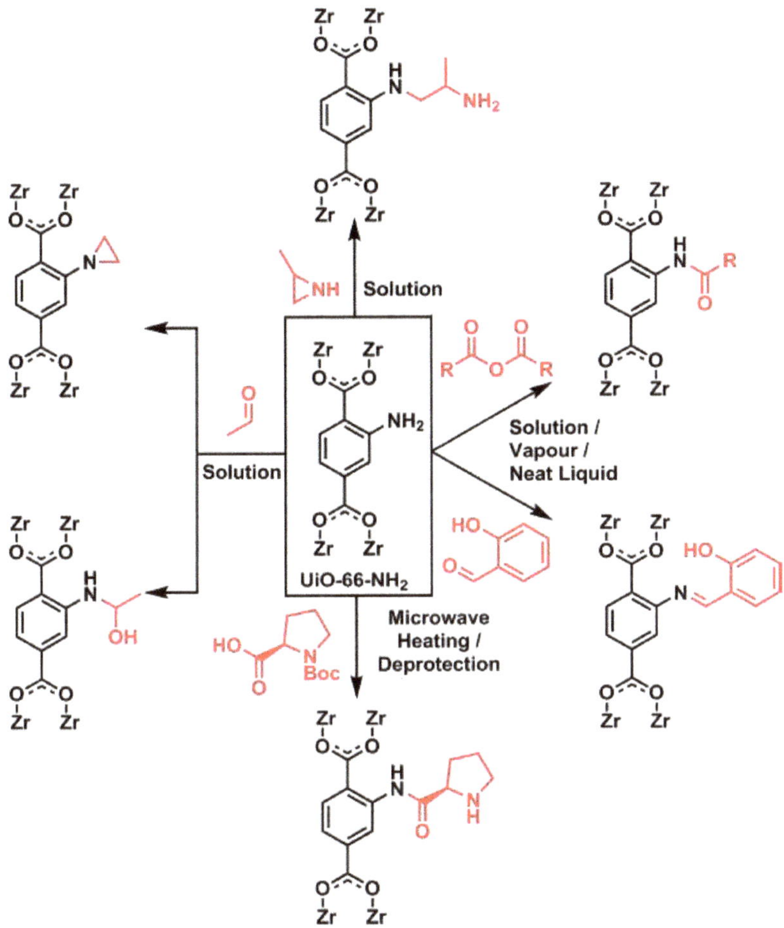

Figure 2.10. Examples of covalent postsynthetic modifications at the pendant amino units of UiO-66-NH_2. (Reproduced from Ref. 21 with permission from the Wiley.)

2.3.3. *Computational screening of MOFs as adsorbents*

As mentioned earlier, enormous amount of MOFs were synthesized so far in the literature and thousands of new MOF structures are possible for synthesis. It is very difficult to synthesize each of the thousands of MOFs and screening them for gas adsorption-based separations is almost impossible. Computational studies helps to understand the gas adsorption behavior of these thousands MOFs in a faster way and screen them with the desired properties (such as adsorption isotherm, diffusivity and also the binding energy) for a specific gas separation application. Recently, high-throughput computational screening of MOFs, especially for adsorption-based gas separations, has gained lot of interests among computational scientists. Wilmer *et al.* paved way to computationally design the metal-organic frameworks and high throughput screening of thousands of possible MOFs for gas separation applications.[27] The performance of these materials is calculated using the computational methods based on the structural properties such as pore size, pore volume and also based on their interaction with gas molecules.[27]

Grand Canonical Monte Carlo (GCMC) methods were used extensively to study the gas adsorption in metal organic frameworks.[28] GCMC methods can simulate the exchange of gas molecules between the porous structure of the MOF and the gas feed i.e. both adsorption and desorption, allowing the simulation of thermodynamic equilibrium of gas adsorption between two phases.[29] Coudert *et al.* summarized the different computational methods used to study the gas adsorption properties of the metal-organic frameworks.[29] Computational analysis was performed on the adsorption separation of CO_2 from flue gases in Cu-BTC MOF to investigate the applicability of MOFs for this application.[30] The GCMC simulation can model not only the adsorption of pure components, but also mixtures of gases by specifying the chemical potential of each component in the mixture. The conventional GCMC simulation was employed to calculate the pure and mixed gas adsorption in Cu-BTC MOF and shows that the electrostatic field inside the pores of the MOF is important for the selective adsorption of a quadrapole gas molecule from mixture of gases.[31] GCMC simulations are limited by the accuracy of a given force field to describe host–guest interactions in the system. These are not very accurate in the case of chemisorption of gas molecules in the open metal center MOFs. *Ab initio* quantum chemistry methods was used, mostly by Density Functional Theory, to analyze the interactions of the gas with unsaturated metal site.[29] Density functional theory (DFT) calculations solves the time-independent Schrödinger equation to find minimum energy structures, binding energies, and details of the electronic structure.[28] The adsorption isotherms that are simulated usually does not match precisely with the experimentally measured isotherms because of the defect formation and presence of guest molecules in the experimentally synthesized MOF samples.

Wilmer *et al.* generated 137,953 hypothetical MOF structures from a library of 102 building blocks that were extracted from real MOFs as shown in Figure 2.11.[27] The pore-size distribution, surface area and methane-storage capacity were calculated for each one of them. Methane uptake was calculated for all the structures using GCMC simulations at

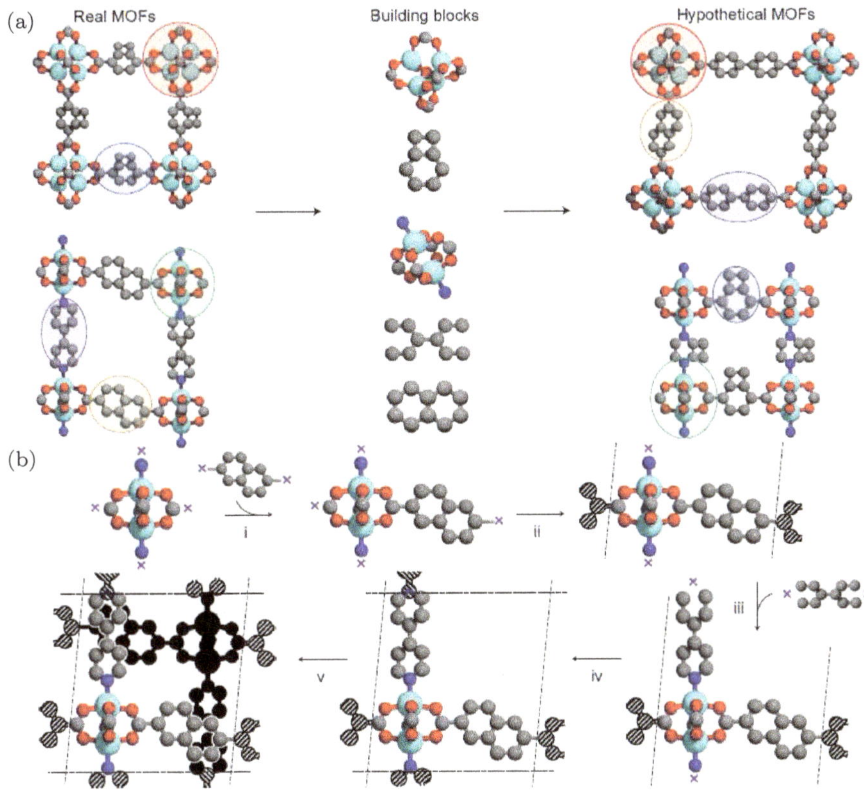

Figure 2.11. Visual summary of the hypothetical MOF-generation strategy. a, Crystal structures of existing MOFs were obtained from X-ray diffraction data (a, left) and subsequently divided into building blocks (a, middle) that then could be recombined to form new, hypothetical MOFs (a, right). b, The recombination process occurs by stepwise addition of building blocks (i), which are attached at their connection sites (purple Xs). Building blocks are also connected across periodic boundaries (ii, hashed circles indicate mirror images). The process repeats (iii to iv) until all connection sites are utilized. An interpenetrated MOF may be generated if enough space exists (v, black circles indicate atoms belonging to one of two interpenetrated frameworks). Grey, red, blue and turquoise spheres represent carbon, oxygen, nitrogen and zinc atoms, respectively. Hydrogen atoms are omitted for clarity. (Reproduced from Ref. 27 with permission from the Springer Nature.)

35 bar and 298 K. It takes huge amount of time to calculate the methane uptake in all those 137,953 hypothertical MOFs and to speed up these calculations, the calculations were performed in stages. Short GCMC simulations were performed for all structures to screen those materials and the top 5% of those MOFs in terms of methane storage were screened with longer simulations. The top 5% MOFs from the second stage were further analyzed for methane storage using even longer simulations. Using this methodology, over 300 hypothetical MOFs shown much higher performance than one of the best known porous materials for methance storage. Changes in the methane adsorption of those thousands of MOFs calculated using GCMC simulations with the pore size of the MOFs was also reported.

Qaio *et al.* extended this high throughput technique for screening all those 137,953 hypothertical MOFs as membranes for the separation of CO_2, N_2 and CH_4.[32] They calculated pore limiting diameter, diffusivity, solubility and permeability of pure gases and mixed gases in a four-stage approach and identified five best MOFs for membrane based gas separation applications. In another study, Qaio and colleagues screened 4,764 computation-ready, experimental metal-organic frameworks (CoRE-MOFs) using molecular simulations for CO_2 separation from flue gas (CO_2/N_2) and natural gas (CO_2/CH_4). Considering adsorption capacity and selectivity, alkali- and alkaline-MOFs has superior performance compared to the other MOFs in the CORe MOFs.[33] One aspect of this study, that is important for industrial applicability, is to consider the regenerability and working capacity of the adsorbent along with the adsorption capacity. The alkali- and alkaline-MOFs that showed high performance in terms of adsorption capacity were not good, considering regenerablity and working capacity. Figure 2.12 shows the three-dimensional relationships of selectivity, regenerability and working capacity and observed that lanthanide-based CoRE-MOFs which has lanthanides as open metal sites showed the highest gas separation performance.

Valenzano *et al.* relied on the computational method to study the gas adsorption behavior of Mg-MOF-74 using a combined methodology comprising variable temperature infrared spectroscopy and ab initio periodic DFT-D calculations using the CRYSTAL code.[34] The standard adsorption enthalpy were also calculated including empirical correction for dispersion forces and adsorption enthalpy of CO, N_2 and CO_2 are −29, −21, and −47 kJ mol-1, respectively which were in agreement with the experimental

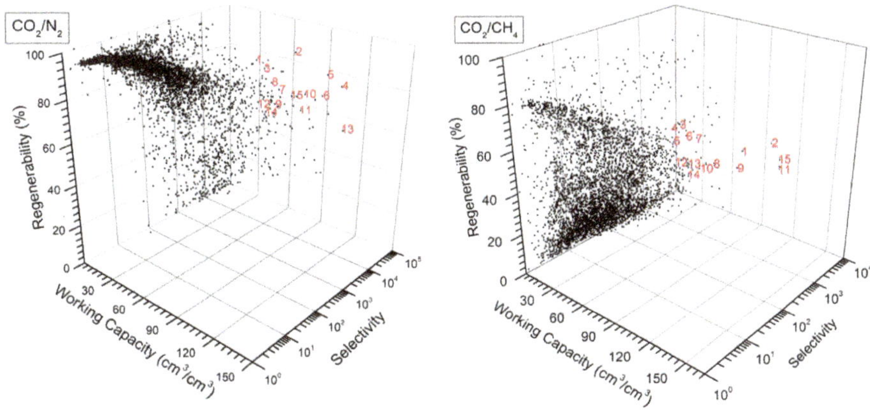

Figure 2.12. Relationships among selectivity S, regenerability R and working capacity ΔN_{CO_2}. (Reproduced from Ref. 33 with permission from the Royal Society of Chemistry.)

results. Molecular simulations were used to predict the adsorption capacity CO_2, CH_4, H_2 and N_2 in an ionic rhozeolite-like MOF and observed very high adsorption selectivity CO_2 gas compared to the other gases.[35]

GCMC can only be used to simulate the adsorption in metal-organic frameworks assuming that they are rigid. As mentioned earlier in this chapter, flexible MOFs are gaining lot of interest in terms of adsorption-based gas separation. Recently, researchers started working on simulating the adsorption behavour in flexible MOFs by performing simulations in a hybrid MC/MD setup or Hybrid Monte Carlo.[36–38] Heinen and Dubbeldam reviewed in detail about the development of hybrid Monte Carlo methods for the simulation of structural changes in the flexible MOFs.[38] Ghoufi *et al.* used osmotic (or hybrid) Monte Carlo simulations to understand the CO_2 adsorption-induced structural transition of MIL-53(Cr) MOF. This osmotic Monte Carlo simulations were able to describe both thermal, mechanical and adsorption-induced structural transitions between the two phases of MIL-53(Cr).[36]

MOFs with coordinatively unsaturated metal sites (open metal center MOFs) showed high gas adsorption properties compared to saturated metal site MOFs. Prediction of gas adsorption behavior of these MOFs using conventional molecular simulations usually results in error. The review by Fischer *et al.* provides more details about the development of simulation methods to study the gas adsorption in open metal center MOFs.[39] They concluded that hybrid approaches that incorporate quantum mechanics-based information into classical models improves the gas adsorption predictions relative to conventional force-fields. Yong and Zhong also used combined GCMC simulation and density functional theory calculation to provide the hydrogen adsorption mechanism in MOF-505.[40] They observed that hydrogen mostly adsorbs on the metal-oxygen clusters in the directions of coordinatively unsaturated open metal sites.

Dzubak *et al.* recently developed a novel computational approach to obtain accurate force fields for MOFs with open metal sites as shown in Figure 2.12.[41] The special feature of this method is based on the nonempirical model potential (NEMO) methodology. These simulations were used to calculate the heat of adsorption, Henry's coefficients, and adsorption isotherms of CO_2 and N_2 in MOF-74(Mg) with a perfect crystalline structure. They found that there is excellent agreement between the simulation results and the experimental data accounting that not all Mg sites are accessible for gas adsorption. They also observed that the simulated pure component adsorption isotherms can be combined with Ideal adsorbed solution theory (IAST) to estimate mixed-gas adsorption behavior in such types of MOFs. IAST predictions based on the simulated single-gas isotherms are in good agreement with the mixed gas isotherms as shown in Figure 2.13.

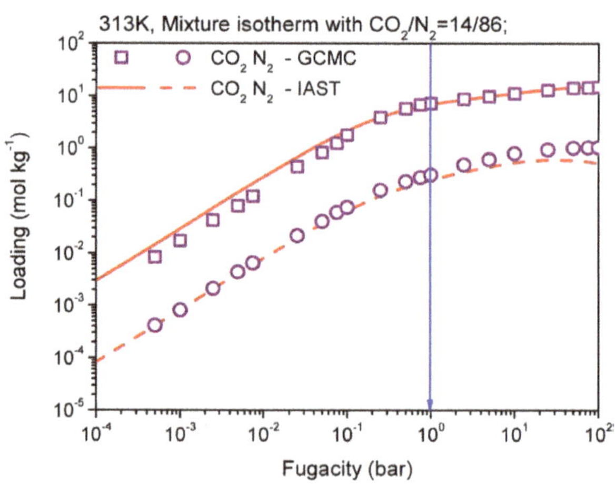

Figure 2.13. Simulated and predicted mixture isotherms. Open symbol shows the simulated CO_2/N_2 (CO_2:N_2 = 14:86 in the bulk phase) mixture isotherms in Mg-MOF-74 at 313 K. Red line shows the mixture isotherm predicted by IAST. Blue arrow indicates the flue gas condition. (Reproduced from Ref. 41 with permission from the Springer Nature.)

2.4. Gas separation applications

Gas separations using adsorbents and membranes gained a lot of interests among academic and industrial researchers to reduce the cost of the gas separation. One of the best materials that has the potential to separate these gases are the metal-organic frameworks. Several MOFs displayed adsorption selectivity towards some gases and considered them for further development in industrial applications. It is also shown that these MOF structures can be modified and functionalized to enhance the desired gas adsorption selectivity. Some of the adsorption-based gas separation applications of MOFs are carbon capture, xenon separation, petrochemical separations, air separation, and removal of toxic gases from process gas. The development of MOFs for these gas separation applications is clearly discussed in this section.

2.4.1. *Carbon dioxide separation*

CO_2 is considered one of the major cause for global warming and carbon capture from the point source emissions has been a great discussion subject in recent years to reduce the greenhouse gas emissions. Most of the gas separation applications studied for MOFs are related to the carbon capture. The three main separation processes are CO_2 separation from coal power plant flue gas (CO_2/N_2), from integrated gasification combined cycle (IGCC, CO_2/H_2) plant, and natural gas (CO_2/CH_4). The review papers by Hong-Cai Zhou group[42] and Jeffrey R. Long group[43] clearly discussed development of MOFs for the above mentioned carbon capture applications. Considered metal-organic frameworks should have high CO_2 adsorption capacity, high selectivity, and low-cost regeneration in order to apply them for CO_2 capture applications. As suggested by Mason *et al.*, the presence of strong CO_2 adsorption sites is crucial to apply the MOFs for post-combustion CO_2 capture using TSA process.[5]

Mg-MOF-74, an open metal center MOF, is one of the most studied MOF for CO_2 capture applications. Yaghi and coworkers showed that Mg-MOF-74 has 8.9 wt% dynamic capacity for CO_2 and mentioned that this MOF has right balance between high adsorption capacity and heat of adsorption.[44] Comparison of the separation properties of Mg-MOF-74 with standard CO_2 capture materials shows that this material has a very high CO_2

adsorption capacity with moderate regeneration conditions. Many research reports aimed at increasing the surface area of the MOFs to improve the CO_2 adsorption capacity for high pressure applications. Furukawa *et al.* synthesized a ultrahigh porosity MOFs, named MOF-200 and MOF-210 which showed the highest BET and Langmuir surface area so far and also showed record high CO_2 adsorption capacity of 54.5 mol/kg at 50 bar and 298 K.[45] This authors also mentioned that the surface areas of MOF-200 and MOF-210 are close to the higher limit that we can get for solid materials and this also might be the highest CO_2 adsorption capacity. A carbonate-based MOF, which has a unsaturated metal site, was studied by Bae *et al.* for CO_2 separation from CH_4 and compared with the saturated metal MOF counterpart.[46] The comparisons of CO_2/CH_4 selectivities between the MOFs with and without open metal sites strongly suggest that open metal sites in a MOF can enhance the separation of polar gases such as CO_2. Yazaydin *et al.* found that Mg/DOBDC and Ni/DOBDC have the highest CO_2 capacities at 0.1 atm and 298 K among the 14 MOFs that they considered and concluded that MOFs with a high density of open metal sites are promising for the separation of CO_2 from low concentration streams.[47] They also found very good correlation between the heat of adsorption and CO_2 adsorption capacity below 1 bar. Xiang *et al.* studied another MOF called UTSA-16 for CO_2 capture applications and this MOF displayed very high CO_2 uptake at ambient conditions.[48] The CO_2 capture capacity of this MOF was compared to the other MOFs as shown in Figure 2.14. The CO_2 adsorption performance of this MOF is comparable to that of MOF-74. The high adsorption capacities are attributed to the suitable pore size and the terminal water molecules being a strong binding site for CO_2.

The pore size of MOFs should be closer to the molecular size of CO_2 to control the adsorption affinity of CO_2 and to obtain the selectivity towards CO_2 gas. Professor Yaghi group precisely controlled the pore size of eight zeolitic imidazole frameworks systematically and studied the CO_2 adsorption behavior as shown in Figure 2.15.[49] The pore diameters were varied incrementally from 3.8 to 13.1 Å. The functionalities decorating the pores of these ZIFs was also varied from polar cyano- and nitro-groups to nonpolar alkyl groups. The CO_2/CH_4 selectivity concludes that ZIFs with polar functionality, ZIF-78 and ZIF-82, has the highest CO_2 adsorption capacity as shown in the figure. Interpenetrated MOFs, another example of small

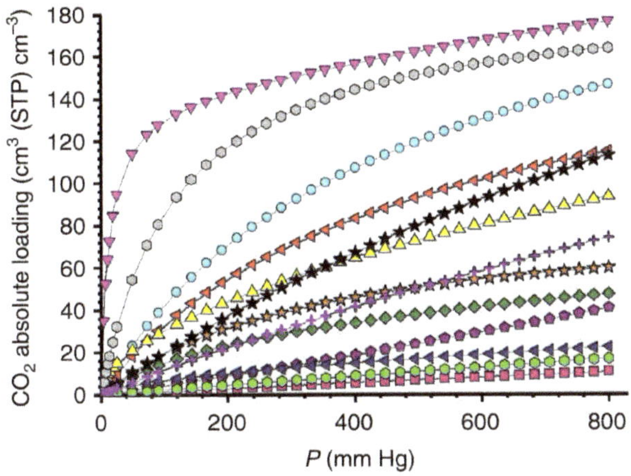

Figure 2.14. Comparison of the absolute loading of CO_2 at 296 K. From top to bottom, MgMOF-74 (magenta down triangle), UTSA-16 (grey hexagon), ZnMOF-74 (cyan circle), bio-MOF-11 (red left triangle), CuBTC (black star), Cu-TDPAT (yellow up triangle), UTSA-20a (violet cross), ZIF-78 (orange star), Zn5(BTA)6(TDA)2 (olive diamond), Zn(bdc)(dabco) (purple pentagon), MIL-101 (navy left triangle), Yb(BPT) (green hexagon) and MOF-177 (pink square). (Reproduced from Ref. 48 with permission from the Springer Nature.)

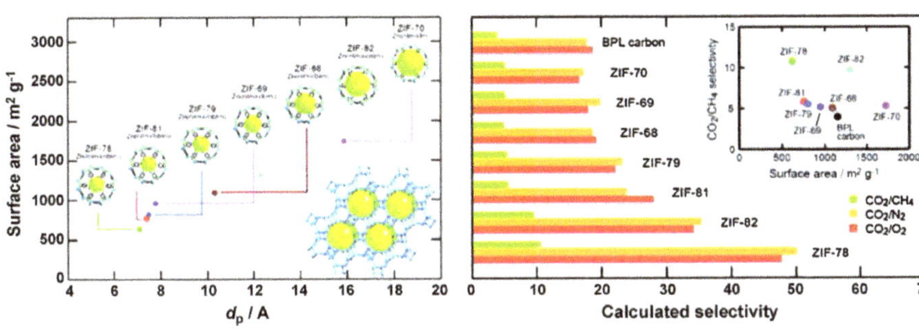

Figure 2.15. Plot of pore diameter (dp) vs surface area for different GME ZIFs (left) and the evaluated CO_2/CH_4, CO_2/N_2, and CO_2/O_2 selectivities in these ZIFs (right). (Reprinted from Ref. 49 with permission from the American Chemical Society.)

controllable pore size MOFs has shown promise for CO_2 separation. The interpenetration or catenation results in reduced pore size and pore volume. The reduction in pore size helped to adsorb more CO_2 because of the availability of more pores for adsorption and changes in the interaction energy with the framework. Interpenetrated SUMOF-2 MOF adsorbed more CO_2 than non-interpenetrated MOF-5 under 273 K and 1 bar since interpenetration increased the electric field gradients.[50]

Amine functionalization of MOFs is one of the most used technique to improve the CO_2 separation performance. Long and coworkers reported the grafting of ethylenediamine in $H_3[(Cu_4Cl)_3(BTTri)_8]$ MOF and hypothesized that one amine is available for CO_2 adsorption while the other amine might attached to the metal center as shown in Figure 2.16.[51] This MOF showed very high heat of adsorption (−90 KJ/mole) at zero coverage because of the weak bond between CO_2 and amines in the MOF framework. Even though heat of sorption is high, this MOF showed very low overall CO_2 adsorption capacity. They later follow up this study with the synthesis of this MOF incorporating N,N-dimethylethylenediamine (mmen) and showed one of the best CO_2 adsorption capacity while having low regeneration temperature.[52] This MOF showed very high heat of adsorption (−96 KJ/mole) at zero coverage because of the strong chemisorption between CO_2 and amines in the MOF framework. An *et al.* synthesized Bio-MOF-11 which contain adenine in its framework and showed high CO_2 adsorption capacity and very high CO_2/N_2 selectivity due to favorable CO_2 adsorption properties

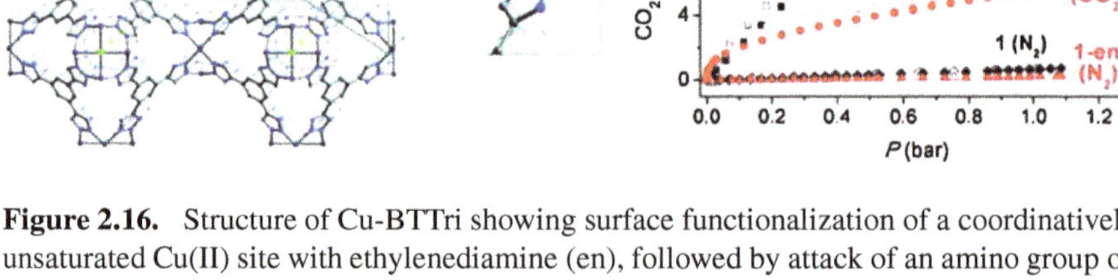

Figure 2.16. Structure of Cu-BTTri showing surface functionalization of a coordinatively unsaturated Cu(II) site with ethylenediamine (en), followed by attack of an amino group on CO_2 (left) and adsorption–desorption isotherms of CO_2 and N_2 at 298 K in Cu-BTTri and Cu-BTTri-en (the inset shows the higher uptake of CO_2 for Cu-BTTri-en compared with Cu-BTTri at low pressures). (Reprinted from Ref. 51 with permission from the American Chemical Society.)

in the presence of Lewis basic amino and pyrimidine groups of adenine along with the narrow pore size distribution.[53] Bae *et al.* grafted the highly polar py-CF_3 groups onto a open metal center MOF, Zn_2(bttb)(CF_3-py)$_2$ which resulted in very high CO_2 uptake and high CO_2/N_2 selectivity at low pressures.[54]

Flexible MOFs, which shows the breathing behavior were also applied for CO_2 capture applications. This flexibility enables the increase in pore volume results in increased of CO_2 adsorption due to the availability of more sorption sites. Bourrelly *et al.* found that CO_2 molecules adsorbed to the hydroxyl groups in MIL-53 (Al) causing the shrinkage in the structure and further increase in CO_2 pressure lead to the reopening of the pore structure (breathing behavior).[55] MIL-53 (Al) showed CO_2 adsorption capacity of 10.4 mol/kg at 30 bar and 304 K, which is well above those of conventional zeolites and comparable with microporous carbons. Substituting Al with Cr in MIL-53 essentially did not cause any change in CO_2 adsorption capacity. Ramsahye *et al.* found that in MIL-53 (Al) and MIL-53(Cr), the adsorption mechanism is mainly governed by the interactions between CO_2 molecules and the μ_2-OH groups.[56] Finsy *et al.* also found that CO_2 strongly interacts with the framework hydroxyl groups in MIL-53 (Al) up to 5 bar while with CO_2/CH_4, selectivity is seven.[57] Further increase in pressure, resulted in the pore opening and the selectivity was reduced to four since the expanded pores are accessible to CH_4. Thallapally *et al.* also synthesized an interpenetrated Zn-based MOF which also shows breathing behavior upon solvent loss and CO_2 adsorption and it has low capacity of 7.1 mol/kg at 30 bar and 298 K.[58] A new flexible MOF, Cu(FMA), synthesized by Li and colleague showed the gate opening mechanism in the presence of CO_2.[46] Interestingly, this MOF showed this flexible behavior to only CO_2, not for CH_4 and N_2. The CO_2 uptake sharply increased to 33.7 $cm^3\ g^{-1}$ at 1 atm from 6.85 $cm^3\ g^{-1}$ at 0.60 atm and resulted in higher CO_2/CH_4 and CO_2/N_2 adsorption selectivity. Kim *et al.* observed similar kind of behavior their MOF, CuTATB-60, by interpenetration. CuTATB-60 showed both high surface area and high CO_2/N_2 selectivity compared to its non-catenated counterpart. This interpenetrated MOF also showed very stable CO_2 adsorption behavior for several cycles and showed very good regeneration ability.[59]

2.4.2. *Xenon separation*

Xenon (Xe) is used for several applications in the field of lighting, electronics, medical and space research. Xenon is a very rare gas available in the air at 87 parts per billion level. It is hard to separate the low concentration Xenon from the air, making it a very costly gas to have and used only in high end applications. Currently, xenon is a byproduct of cryogenic distillation of air and is a very energy-intensive process. The product gas after the separation of other gases contains about 20% Xenon and 80% Krypton. Research on separation of Xe using membranes and adsorption gained a lot of interests to reduce the cost of Xe production. Especially, separation of Xe using metal-organic frameworks as adsorbents has the potential for commercial applications. The purity levels of Xenon needed to use in the above mentioned application is very high (>99%), which means pure Krypton gas can also be produced in other stream. The pure Kr gas can also be used in fluorescent light bulb applications, resonance imaging, photographic flashes, heat trap and others. Thallapally and colleagues published in-depth review on the separation of Xe using MOFs as adsorbents.[60] MOFs can be used as adsorbent to separate the Xe gas and also used as a storage media. Different MOFs with different structural and chemical features are needed for adsorption and storage applications. As suggested by Banerjee *et al.* in their review, in large surface area and high void fraction, functionalized MOFs are good for storage applications and optimal pore size MOFs to fit a Xe atom can be used for the xenon separation.

Mueller and team are one of the first teams to study the adsorption of xenon in MOFs.[61] It was found in this study that the higher amount (about 4 times) of Xe gas can be filled in a gas cylinder at lower pressure using MOF-5 compared to the Xe gas cylinders without any MOF-5 as shown in Figure 2.17. Other inert gases like Krypton and Argon behaved similarly, but to a less extent. This observation lead to the application of MOF-5 as adsorbents in the separation of Xenon from other gases. Muller *et al.* measured that Cu-MOF can adsorb about 60 wt% of Xe gas which was almost two times higher than the activated carbon at similar conditions. Also, faster gas kinetics in MOFs compared to other porous materials makes them great materials to reduce the cost of Xe separation by faster swing operations.

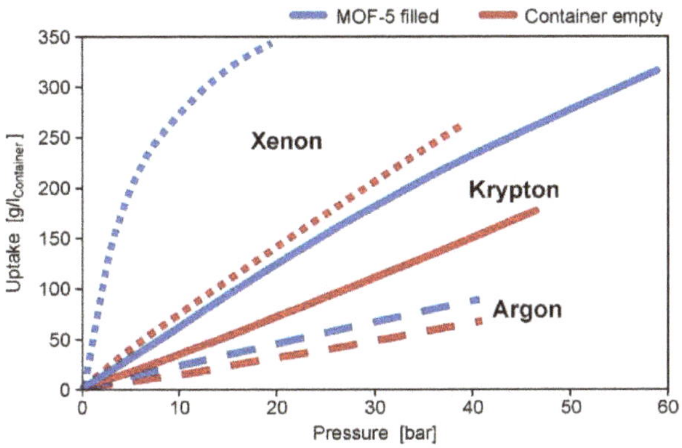

Figure 2.17. Compression curves of rare gases Ar, Kr, Xe and comparison over inflation curves into MOF-5-filled gas containers (room temperature, up to 60 bar; lecture bottles). (Reproduced from Ref. 61 with permission from the Royal Society of Chemistry.)

As mentioned earlier in this chapter, heat of adsorption is an important parameter to be studied for adsorption-based gas separation applications. Farrusseng *et al.* measured the heat of sorption of Xe in IRMOF-1 and IRMOF-3 by using a pulse response experiment in an ultrahigh vacuum reactor.[62] The heat of adsorption of IRMOF-1 and IRMOF-3 are $\Delta 16$ kJ/mol and $\Delta 24$ kJ/mol, respectively which are significantly higher than the Kr. IRMOF-3 showed higher heat of sorption compared to IRMOF-1 due to the Xe polarizability in the presence of -NH_2 in IRMOF-3. The polarizability of Xenon in the presence of other functional groups in IRMOF-1 was studied by Meek *et al.* IRMOF-1 was functionalized with different halogens (-F, -Cl, -Br, -I) and studied the adsorption of N_2, Ar, Kr, and Xe gas using a combination of experimental gas adsorption and molecular simulation.[63] The results showed that the gas adsorption increased with increase in linker polarizability (I $>$ Br $>$ Cl $>$ F) due to the increased in adsorbate interaction with increased linker polarizability. The increased polarizability also resulted in higher Kr/N_2, Xe/N_2, and Xe/Kr selectivity. All the gases showed linear isotherms suggests that adsorption process in this IRMOF-1 is governed by linker polarizability.

Lee *et al.* studied the separation of Xe/Kr mixtures using three different hydrothermally stable MOFs (MIL-100(Fe), MIL-101(Cr), and UiO-66(Zr)).[64] Adsorption isotherms of Xe and Kr in these three MOFs is shown in Figure 2.18. Among those three MOFs, it was observed that UiO-66 MOF

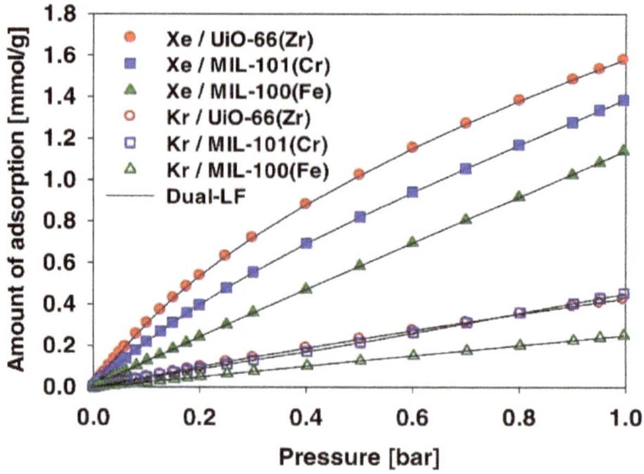

Figure 2.18. Adsorption isotherms of Xe and Kr in three MOFs at 303 K. All the isotherms were fit to the dual-site Langmuir-Freundlich equation. (Reprinted from Ref. 64 with permission from Elsevier.)

is the best candidate in terms of heat of adsorption and selectivity. UiO-66 showed Xe adsorption capacity of 1.58 mmol/g at 303 K and 1 bar with Xe/Kr selectivity of around 4. Large Xe uptake and high Xe/Kr selectivities of UiO-66(Zr) compared to MIL-100(Fe) and MIL-101(Cr) were attributed to the optimum pore size and pore geometry of UiO-66(Zr), which leads to effective interactions between adsorbed atoms and the pore surface.

Several reports also used open metal center MOFs for the separation of Xenon from Krypton gas. Thallapalli and coworkers studied the Xe adsorption in NiDOBDC MOF at noncryogenic temperatures and found that Xe adsorption is higher than MOF-5.[65] The selectivity of the MOF for Xe over Kr is higher compared to activated carbon while having similar selectivity. As a noble gas atom, the only mechanism for xenon to interact is based on its polarizability. Metal cations are very polarizing, which may enhance the interaction with polarizable adsorbents. Allendorf and colleagues measured the adsorption of noble gases in open metal center MOFs using volumetric gas porosimetry and GCMC simulation.[66] Different MOFs were studied systematically to gain knowledge about the effect of metal ion, effect of pore size and topology. The noble gas adsorption behavior of MOF-74 with different metal ions Mg, Co, Ni, and Zn, was studied and observed that the density of open metal sites and their accessibility enhance the interaction with the noble gases. But the type of metal is relatively not important since all of the metal ions studied here showed similar heat of adsorption for the noble gases. nbo series MOFs which have the same metal center was studied to analyze the effect of pore size on the adsorption of noble gases and observed that both gas uptake and Henry's constant increases with decreasing pore size. It is also observed that the effect of gas polarizability on henry's constant becomes more profound as the pore size decreases. Elsaidi *et al.* studied the small pore MOFs, SIFSIX-3-Ni and SIFSIX-3-Fe, for the Xe separation and observed the changes in the adsorption behavior with slight changes in those MOFs.[67] The changes in adsorption behavior with slight changes in the MOF structure attributed the inflection to a disordered transition to ordered transition of the rotational configurations of the pyrazine rings induced by sorbent-sorbate interactions.

2.4.3. *Hydrocarbon separations*

Th separation of light hydrocarbons such as ethane/ethylene, propane/propylene gained a lot of interests these days to reduce cost of ethylene or propylene production compared to the current cryogenic distillation. Adsorbents can be used to reduce the load on the distillation column to reduce olefin/paraffin production cost. It is very difficult to separate the hydrocarbon mixture especially olefin/paraffin due to the similar physical and chemical properties of these molecules. MOFs has been studied as adsorbents by several groups for hydrocarbon separations. The adsorption capacity of olefins in the MOFs is generally high compared to paraffins because of the relatively diffuse p-orbital.

Lamia *et al.* studied the potential HKUST-1 MOF for olefin/paraffin separation.[68] Comparison of experimental and GCMC simulation adsorption isotherms showed that propylene adsorbs on the open metal center due to the specific interactions between the electron-rich p-bonding orbital in propene and the vacant s-orbital of the open metal centers. Yoon *et al.* showed high propylene selectivity and attributed to unsaturated Fe^{2+} and Fe^{3+}sites in MIL-100(Fe).[69] Heat of adsorption of propylene gas is around 70 $kJmol^{-1}$ at low coverage. This high selectivity was observed at very low pressures, and it decreased markedly with increasing pressure, perhaps due to open metal center saturation with the propylene gas. Bae *et al.* studied the gas adsorption behavior for a series of isostructural frameworks M-MOF-74 (M=Co,

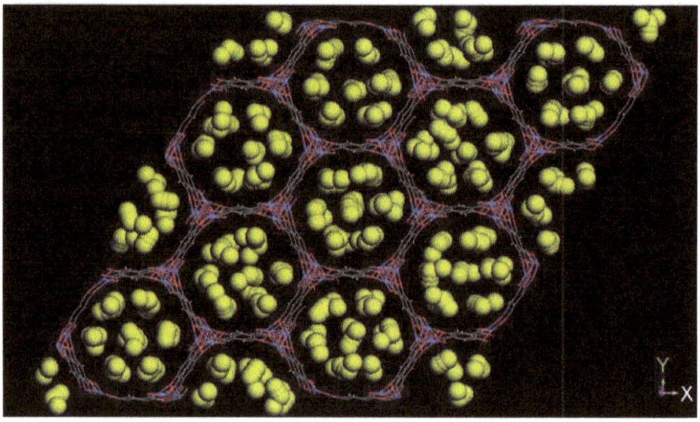

Figure 2.19. A snapshot from a GCMC simulation for C_3H_6 adsorption in Co-MOF-74 at 1 bar and 298 K, looking down the 1-D hexagonal channels. Under this condition, each metal site is occupied by one propene molecule (1.08 C_3H_6 per Co^{2+}). C: gray; H: white; O: red; Co: blue. Yellow spheres indicate the carbon atoms in propene. (Reproduced from Ref. 70 with permission from the Wiley.)

Mn, and Mg) with high concentrations of open metal sites.[70] Adsorption experiments and simulations showed that Co-MOF-74 exhibits the highest thermodynamic propene/propane selectivity ever reported for MOFs, due to strong p-complexation between the open Co^{2+} sites and the propene molecules as shown in Figure 2.19. Geier and colleagues extended this study of M-MOF-74 with other metals like Fe, Ni and Zinc and results showed good performance characteristics, Fe_2(dobdc) and Mn_2(dobdc) exhibit the highest selectivities for the separation of ethylene–ethane and propylene–propane mixtures, respectively.[71] Belarbi and collegues were also studied the adsorption of several hydrocarbons (saturated, unsaturated and cyclic hydrocarbons) in MIL-101(Cr) using gas chromatography equipped with a filled column of the MOF.[72] This study showed that linear hydrocarbons are eluted firstly in the case of the n-hexane isomers series. This work also suggested that creating high densities of open metal sites in pores of limited volume may be a powerful general strategy for generating new adsorbents for olefin/paraffin and other separations. MOFs with Cu^+ or Ag^+ metal centers showed very high selectivity for the adsorption of olefins over paraffins. These metal center showed very high heat of adsorption values (about 100 KJ/mole) because of the strong covalent interactions. Even though they show very high propylene selectivity, these MOF might need very high temperature to desorb these gases.[72]

The feasibility of hydrocarbons separation using flexible MOFs was also studied. The iron MIL-53 showed a closed structure in the dry state and shows more complex adsorption isotherms with multisteps that depend on the nature of the alkane.[73] This complex multistep adsorption behavior in MIL-53 (Fe) attributed to the existence of four discrete pore openings during the whole adsorption process. XRPD patterns and/or the isotherms revealed the changes in structural pieces and the pressure at which these structural transitions occurs decreases as the size of the hydrocarbon increases as shown in Figure 2.20. Nijem *et al.* showed unexpected hydrocarbon selectivity in a flexible MOFs based on the differences in their gate opening pressure and the selectivity depends on both chain length and specific framework-gas interaction.[74] Authors studied the separation performance of C_1-C_4 paraffins and two pairs of C_2 isomers (C_2H_2-C_2H_4 and C_2H_4-C_2H_6), in a flexible framework, RPM3-Zn. Raman spectroscopy and *ab initio* DFT calculations showed that H-bonding between their terminal groups and the C=O bond of bpdc ligand of the framework are the dominant factors in high hydrocarbon

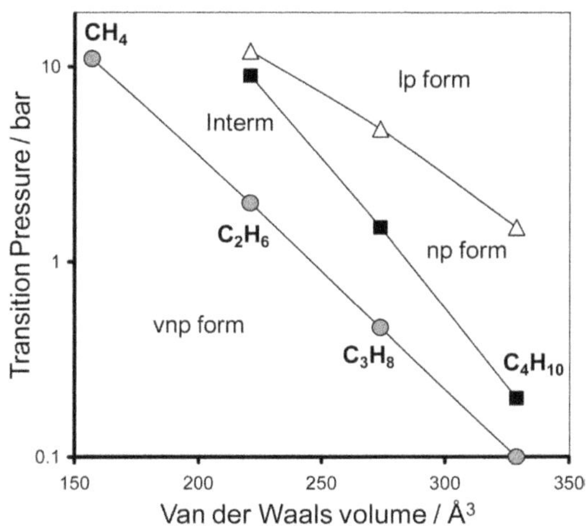

Figure 2.20. Variation of the transition pressures of MIL-53(Fe) at 303 K asa function of the van der Waals volumes of the alkane molecules. (Reprinted from Ref. 73 with permission from the American Chemical Society.)

selectivity and the stronger hydrogen bonding leads to a lower gate opening pressure.

2.5. Conclusions

Because of high porosity, diverse functionalities, and easy modification, metal-organic frameworks (MOFs) have important applications in adsorption. MOFs have shown great application potential as gas separation adsorbents. Most widely studied application of MOFs are CO_2 separation, Nobel gas separation, air separation, hydrocarbon separation and separation of harmful gases. Different kind of MOFs, such as open metal center MOFs, interpenetrated MOF, amine appended MOF and flexible MOFs, were used as adsorbent for gas separations. Apart from measuring the pore size, pore volume and surface area, these MOFs were characterized by developing the adsorption isotherms at different pressures and temperatures. Further research of the MOF development for adsorption-based gas separations needs to focus on designing the MOF for a specific application and testing them with gas mixture. Most importantly, testing these materials under the influence of contaminants, such as water, sulfur and nitrogen compounds, in the process gas should be the focus of the research.

References

1. Li, J.-R., Sculley, J. & Zhou, H.-C. Metal-organic frameworks for separations. *Chem. Rev.* **112**, 869–932 (2012).
2. Wiessner, F. G. Basics and industrial applications of pressure swing adsorption (PSA), the modern way to separate gas. *Gas Separation & Purification*, **2**, 115–119 (1988).
3. Hu, Z., Wang, Y., Shah, B. B. & Zhao, D. CO_2 capture in metal-organic framework adsorbents: An engineering perspective. *Adv. Sustain. Syst.* **3**, 1800080 (2019).
4. Moellmer, J. *et al.* Insights on adsorption characterization of metal-organic frameworks: A benchmark study on the novel soc-MOF. *Microporous Mesoporous Mater.* **129**, 345–353 (2010).
5. Mason, J. A., Sumida, K., Herm, Z. R., Krishna, R. & Long, J. R. Evaluating metal-organic frameworks for post-combustion carbon dioxide capture via temperature swing adsorption. *Energy Environ. Sci.* **4**, 3030 (2011).
6. Walton, K. S. & Sholl, D. S. Predicting multicomponent adsorption: 50 years of the ideal adsorbed solution theory. *AIChE J.* **61**, 2757–2762 (2015).
7. Liu, J., Tian, J., Thallapally, P. K. & McGrail, B. P. Selective CO_2 capture from flue gas using metal-organic frameworks — A fixed bed study. *J. Phys. Chem. C* **116**, 9575–9581 (2012).

8. Lillo-Ródenas, M. A., Fletcher, A. J., Thomas, K. M., Cazorla-Amorós, D. & Linares-Solano, A. Competitive adsorption of a benzene–toluene mixture on activated carbons at low concentration. *Carbon N. Y.* **44**, 1455–1463 (2006).
9. Liu, Q. *et al.* Adsorption of carbon dioxide by MIL-101(Cr): Regeneration conditions and influence of flue gas contaminants. *Sci. Rep.* **3**, 2916 (2013).
10. Meng, W. *et al.* Tuning expanded pores in metal-organic frameworks for selective capture and catalytic conversion of carbon dioxide. *ChemSusChem* **11**, 3751–3757 (2018).
11. Shekhah, O. *et al.* Made-to-order metal-organic frameworks for trace carbon dioxide removal and air capture. *Nat. Commun.* **5**, 4228 (2014).
12. Schneemann, A. *et al.* Flexible metal-organic frameworks. *Chem. Soc. Rev.* **43**, 6062–6096 (2014).
13. Kitagawa, S., Kitaura, R. & Noro, S. Functional porous coordination polymers. *Angew. Chemie Int. Ed.* **43**, 2334–2375 (2004).
14. Bureekaew, S., Shimomura, S. & Kitagawa, S. Chemistry and application of flexible porous coordination polymers. *Sci. Technol. Adv. Mater.* **9**, 014108 (2008).
15. Rogge, S. M. J. *et al.* Modeling gas adsorption in flexible metal-organic frameworks via hybrid Monte Carlo/molecular dynamics schemes. *Adv. Theory Simulations* **2**, 1800177 (2019).
16. Loiseau, T. *et al.* A rationale for the large breathing of the porous Aluminum Terephthalate (MIL-53) upon hydration. *Chem. — A Eur. J.* **10**, 1373–1382 (2004).
17. Jiang, H.-L., Makal, T. A. & Zhou, H.-C. Interpenetration control in metal-organic frameworks for functional applications. *Coord. Chem. Rev.* **257**, 2232–2249 (2013).
18. Kuang, X. *et al.* Assembly of a metal-organic framework by sextuple intercatenation of discrete adamantane-like cages. *Nat. Chem.* **2**, 461–465 (2010).
19. Tanabe, K. K. & Cohen, S. M. Postsynthetic modification of metal-organic frameworks — a progress report. *Chem. Soc. Rev.* **40**, 498–519 (2011).
20. Wang, Z. & Cohen, S. M. Postsynthetic modification of metal-organic frameworks. *Chem. Soc. Rev.* **38**, 1315 (2009).
21. Marshall, R. J. & Forgan, R. S. Postsynthetic modification of zirconium metal-organic frameworks. *Eur. J. Inorg. Chem.* **2016**, 4310–4331 (2016).
22. Nguyen, J. G. & Cohen, S. M. Moisture-resistant and superhydrophobic metal-organic frameworks obtained via postsynthetic modification. *J. Am. Chem. Soc.* **132**, 4560–4561 (2010).
23. Wang, Z., Tanabe, K. & Cohen, S. Tuning. Hydrogen sorption properties of metal-organic frameworks by postsynthetic covalent modification. *Chem. — A Eur. J.* **16**, 212–217 (2010).
24. Steven S. Kaye, Anne Dailly, Omar M. Yaghi & Jeffrey R. Long. Impact of Preparation and Handling on the Hydrogen Storage Properties of Zn_4O(1,4-benzenedicarboxylate)$_3$ (MOF-5). (2007). doi:10.1021/JA076877G
25. Venna, S. R. *et al.* Fabrication of MMMs with improved gas separation properties using externally-functionalized MOF particles. *J. Mater. Chem. A* **3**, 5014–5022 (2015).
26. Demessence, A., D'Alessandro, D. M., Foo, M. L. & Long, J. R. Strong CO_2 binding in a water-stable, Triazolate-bridged metal-organic framework functionalized with Ethylenediamine. *J. Am. Chem. Soc.* **131**, 8784–8786 (2009).

27. Wilmer, C. E. *et al.* Large-scale screening of hypothetical metal-organic frameworks. *Nat. Chem.* **4**, 83–89 (2012).
28. Colón, Y. J. & Snurr, R. Q. High-throughput computational screening of metal-organic frameworks. *Chem. Soc. Rev.* **43**, 5735–5749 (2014).
29. Coudert, F.-X. & Fuchs, A. H. Computational characterization and prediction of metal-organic framework properties. *Coord. Chem. Rev.* **307**, 211–236 (2016).
30. Yang, Q., Xue, C., Zhong, C. & Chen, J.-F. Molecular simulation of separation of CO_2 from flue gases in CU-BTC metal-organic framework. *AIChE J.* **53**, 2832–2840 (2007).
31. Sun, B., Kayal, S. & Chakraborty, A. Study of HKUST (Copper benzene-1,3,5-tricarboxylate, Cu-BTC MOF)-1 metal organic frameworks for CH4 adsorption: An experimental Investigation with GCMC (grand canonical Monte-carlo) simulation. *Energy* **76**, 419–427 (2014).
32. Qiao, Z., Peng, C., Zhou, J. & Jiang, J. High-throughput computational screening of 137953 metal-organic frameworks for membrane separation of a $CO_2/N_2/CH_4$ mixture. *J. Mater. Chem. A* **4**, 15904–15912 (2016).
33. Qiao, Z., Zhang, K. & Jiang, J. In silico screening of 4764 computation-ready, experimental metal-organic frameworks for CO_2 separation. *J. Mater. Chem. A* **4**, 2105–2114 (2016).
34. Valenzano, L. *et al.* Computational and experimental studies on the adsorption of CO, N_2, and CO_2 on Mg-MOF-74. *J. Phys. Chem. C* **114**, 11185–11191 (2010).
35. Babarao, R. & Jiang, J. Unprecedentedly high selective adsorption of gas mixtures in *rho* Zeolite-like metal-organic framework: A molecular simulation study. *J. Am. Chem. Soc.* **131**, 11417–11425 (2009).
36. Ghoufi, A. & Maurin, G. Hybrid Monte Carlo simulations combined with a phase mixture model to predict the structural transitions of a porous metal-organic framework material upon adsorption of guest molecules. *J. Phys. Chem. C* **114**, 6496–6502 (2010).
37. Rogge, S. M. J. *et al.* Modeling gas adsorption in flexible metal-organic frameworks via hybrid Monte Carlo/molecular dynamics schemes. *Adv. Theory Simulations* **2**, 1800177 (2019).
38. Heinen, J. & Dubbeldam, D. On flexible force fields for metal-organic frameworks: Recent developments and future prospects. *Wiley Interdiscip. Rev. Comput. Mol. Sci.* **8**, e1363 (2018).
39. Fischer, M., Gomes, J. R. B. & Jorge, M. Computational approaches to study adsorption in MOFs with unsaturated metal sites. *Mol. Simul.* **40**, 537–556 (2014).
40. Yang, Q. & Zhong, C. Understanding Hydrogen Adsorption in Metal-Organic Frameworks with Open Metal Sites: A Computational Study. (2005). doi:10.1021/JP055908W
41. Dzubak, A. L. *et al.* Ab initio carbon capture in open-site metal-organic frameworks. *Nat. Chem.* **4**, 810–816 (2012).
42. Li, J.-R. *et al.* Carbon dioxide capture-related gas adsorption and separation in metal-organic frameworks. *Coord. Chem. Rev.* **255**, 1791–1823 (2011).
43. Sumida, K. *et al.* Carbon dioxide capture in metal-organic frameworks. *Chem. Rev.* **112**, 724–781 (2012).

44. Britt, D., Furukawa, H., Wang, B., Glover, T. G. & Yaghi, O. M. Highly efficient separation of carbon dioxide by a metal-organic framework replete with open metal sites. *Proc. Natl. Acad. Sci. U. S. A.* **106**, 20637–20640 (2009).
45. Furukawa, H. *et al.* Ultrahigh porosity in metal-organic frameworks. *Science* **329**, 424–428 (2010).
46. Li, B. & Chen, B. A flexible metal-organic framework with double interpenetration for highly selective CO_2 capture at room temperature. *Sci. China Chem.* **59**, 965–969 (2016).
47. Yazaydın, A. O. *et al.* Screening of metal-organic frameworks for carbon dioxide capture from flue gas using a combined experimental and modeling approach. *J. Am. Chem. Soc.* **131**, 18198–18199 (2009).
48. Xiang, S. *et al.* Microporous metal-organic framework with potential for carbon dioxide capture at ambient conditions. *Nat. Commun.* **3**, 954 (2012).
49. Banerjee, R. *et al.* Control of pore size and functionality in isoreticular zeolitic imidazolate frameworks and their carbon dioxide selective capture properties. *J. Am. Chem. Soc.* **131**, 3875–3877 (2009).
50. Yao, Q. *et al.* Interpenetrated metal-organic frameworks and their uptake of CO_2 at relatively low pressures. *J. Mater. Chem.* **22**, 10345 (2012).
51. Demessence, A., D'Alessandro, D. M., Foo, M. L. & Long, J. R. Strong CO_2 binding in a water-stable, triazolate-bridged metal-organic framework functionalized with ethylenediamine. *J. Am. Chem. Soc.* **131**, 8784–8786 (2009).
52. McDonald, T. M., D'Alessandro, D. M., Krishna, R. & Long, J. R. Enhanced carbon dioxide capture upon incorporation of N,N′-dimethylethylenediamine in the metal-organic framework CuBTTri. *Chem. Sci.* **2**, 2022 (2011).
53. An, J., Geib, S. J. & Rosi, N. L. High and selective CO_2 uptake in a cobalt adeninate metal-organic framework exhibiting pyrimidine- and amino-decorated pores. *J. Am. Chem. Soc.* **132**, 38–39 (2010).
54. Bae, Y.-S., Farha, O. K., Hupp, J. T. & Snurr, R. Q. Enhancement of CO_2/N_2 selectivity in a metal-organic framework by cavity modification. *J. Mater. Chem.* **19**, 2131 (2009).
55. Sandrine Bourrelly, *et al.* Different Adsorption Behaviors of Methane and Carbon Dioxide in the Isotypic Nanoporous Metal Terephthalates MIL-53 and MIL-47. (2005). doi:10.1021/JA054668V
56. Ramsahye, N. A., *et al.* Probing the Adsorption Sites for CO_2 in Metal Organic Frameworks Materials MIL-53 (Al, Cr) and MIL-47 (V) by Density Functional Theory. (2007). doi:10.1021/JP075782Y
57. Finsy, V. *et al.* Separation of CO_2/CH_4 mixtures with the MIL-53(Al) metal-organic framework. *Microporous Mesoporous Mater.* **120**, 221–227 (2009).
58. Thallapally, P. K. *et al.* Flexible (Breathing) interpenetrated metal-organic frameworks for CO_2 separation applications. *J. Am. Chem. Soc.* **130**, 16842–16843 (2008).
59. Kim, J. *et al.* Control of catenation in CuTATB-n metal-organic frameworks by sonochemical synthesis and its effect on CO_2 adsorption. *J. Mater. Chem.* **21**, 3070 (2011).
60. Banerjee, D., Simon, C. M., Elsaidi, S. K., Haranczyk, M. & Thallapally, P. K. Xenon gas separation and storage using metal-organic frameworks. *Chem* **4**, 650 (2018).
61. Mueller, U. *et al.* Metal-organic frameworks — prospective industrial applications. *J. Mater. Chem.* **16**, 626–636 (2006).

62. Farrusseng, D. *et al.* Heats of adsorption for seven gases in three metal-organic frameworks: Systematic comparison of experiment and simulation. *Langmuir* **25**, 7383–7388 (2009).
63. Meek, S. T., Teich-McGoldrick, S. L., Perry, J. J., Greathouse, J. A. & Allendorf, M. D. Effects of polarizability on the adsorption of noble gases at low pressures in monohalogenated isoreticular metal-organic frameworks. *J. Phys. Chem. C* **116**, 19765–19772 (2012).
64. Lee, S.-J. *et al.* Adsorptive separation of xenon/krypton mixtures using a zirconium-based metal-organic framework with high hydrothermal and radioactive stabilities. *J. Hazard. Mater.* **320**, 513–520 (2016).
65. Thallapally, P. K., Grate, J. W. & Motkuri, R. K. Facile xenon capture and release at room temperature using a metal-organic framework: A comparison with activated charcoal. *Chem. Commun.* **48**, 347–349 (2012).
66. Perry, J. J. *et al.* Noble gas adsorption in metal-organic frameworks containing open metal sites. *J. Phys. Chem. C* **118**, 11685–11698 (2014).
67. Elsaidi, S. K. *et al.* Effect of ring rotation upon gas adsorption in SIFSIX-3-M (M = Fe, Ni) pillared square grid networks. *Chem. Sci.* **8**, 2373–2380 (2017).
68. Lamia, N. *et al.* Adsorption of propane, propylene and isobutane on a metal-organic framework: Molecular simulation and experiment. *Chem. Eng. Sci.* **64**, 3246–3259 (2009).
69. Yoon, J. W. *et al.* Controlled reducibility of a metal-organic framework with coordinatively unsaturated sites for preferential gas sorption. *Angew. Chemie Int. Ed.* **49**, 5949–5952 (2010).
70. Bae, Y.-S. *et al.* High propene/propane selectivity in isostructural metal-organic frameworks with high densities of open metal sites. *Angew. Chemie Int. Ed.* **51**, 1857–1860 (2012).
71. Geier, S. J. *et al.* Selective adsorption of ethylene over ethane and propylene over propane in the metal-organic frameworks M_2(dobdc) (M = Mg, Mn, Fe, Co, Ni, Zn). *Chem. Sci.* **4**, 2054 (2013).
72. Belarbi, H. *et al.* Adsorption and separation of hydrocarbons by the metal organic framework MIL-101(Cr). *Colloids Surfaces A Physicochem. Eng. Asp.* **520**, 46–52 (2017).
73. Llewellyn, P. L. *et al.* Complex adsorption of short linear alkanes in the flexible metal-organic-framework MIL-53(Fe). *J. Am. Chem. Soc.* **131**, 13002–13008 (2009).
74. Nijem, N. *et al.* Tuning the gate opening pressure of metal-organic frameworks (MOFs) for the selective separation of hydrocarbons. *J. Am. Chem. Soc.* **134**, 15201–15204 (2012).

Chapter 3

Metal-Organic Frameworks as Appealing Materials for Gas Separations

3.1. Introduction

MOFs have been studied for several applications such as gas and liquid separations, gas storage, catalysis, sensing, ion exchange, and drug delivery because of their highly desirable properties, such as uniform micropores, high surface areas, chemical functional tunability and exceptional thermal and chemical stability.[1] Gas separations using MOFs have been studied extensively by several researchers compared to other applications. Also, several companies synthesized MOFs for large scale gas separation applications. Their high surface areas and open sponge-like structures make them ideal for separating mixtures of gases or for storing large volumes of gas at low pressures. MOF structures can be designed for specific applications and can be synthesized using several metal nodes and organic linkers with desired properties specifically for gas separation applications. Other properties such as remarkable high gas uptakes, large accessible pore volumes, and pore sizes in the range of the kinetic diameters of gases makes MOFs as appealing materials for gas separation applications.[2]

There are several published reviews discussing the properties of MOFs and their applications for gas separation.[3–8] Yaghi and coworkers summarized the crucial developments in MOF chemistry over the years that made them MOFs as promising candidates for gas storage and gas separation applications as adsorbents and membranes.[9] The rational design of MOFs using reticular chemistry to vary the chemical and physical properties by changing the functional groups on the metal salts and organic linkers and most importantly, changing their pore arrangements. Other developments include the molecular-level control and atomically precise chemical transformations of MOFs, ultrahigh porosity and high surface areas and

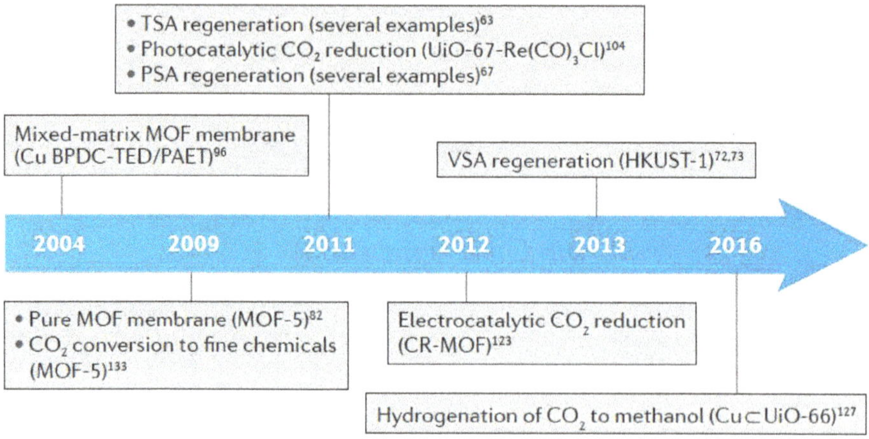

Figure 3.1. Chronology of key achievements made in the field of MOFs and CO_2. Over the past two decades, significant discoveries have been made (and continue to be made) with respect to applying metal-organic frameworks (MOFs) to various components of the CO_2 cycle. PSA, pressure swing adsorption; TSA, temperature swing adsorption; VSA, vacuum swing adsorption; BPDC, biphenyl-4,4′-dicarboxylate; PAET, poly(3-acetoxyethylthiophene); TED, triethylenedi amine. (Reproduced from Ref. 9 with permission from the Springer Nature.)

multivariant MOFs. MOFs can be designed to have coordinatively unsaturated metal sites, specific heteroatoms, covalent functionalization, other building unit interactions, porosity, and high surface area with a level of precision that are needed for specific gas separation applications. CO_2 separation is most studied in gas separation application using MOFs by applying several methods such as adsorption, membranes, catalytic reduction and chemical conversion. The progress of the usage of MOFs for CO_2 separation application is shown chronologically in Figure 3.1 by Yaghi and coworkers.[9] Here, the properties that made these MOFs applicable for gas separation are discussed.

3.2. Gas separation properties of MOFs

3.2.1. *Surface area and pore size properties of the MOFs*

Surface area and pore sizes are one of the most important properties compared to other properties of MOFs in order to apply them for gas separation applications. The choice of precursors along with the synthesis conditions dictates the structure and properties of the MOFs. The coordination preference of the metal influences the size and shape of pores by dictating the number of ligands that can bind to the metal and the orientation of ligands.[10] Most importantly, the process of removal of guest molecules from the MOF pores dictates the surface area of the MOFs.[11] As mentioned in Chapter 2, these guest molecules should be removed from the MOF pores without collapsing the pores in order to obtain the high surface area and open pore volume.

Figure 3.2 shows the surface area of the several MOFs in comparison with the other porous materials.[1] The surface area of MOFs were improved tremendously over the years from ~2000 m^2/g to ~7000 m^2/g. In 2012, Farha *et al.* reported MOFs (NU-109 and NU-110) that showed BET surface area of more than 7000 m^2/g.[12] The surface area of one gram of this MOF is more than the surface area of one football field. So far this is

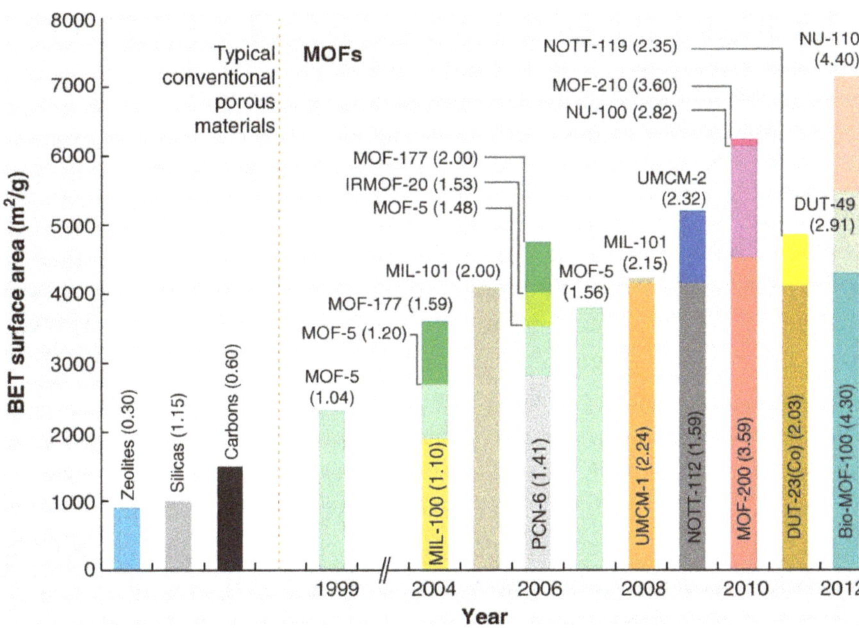

Figure 3.2. Progress in the synthesis of ultrahigh-porosity MOFs. BET surface areas of MOFs and typical conventional materials were estimated from gas adsorption measurements. The values in parentheses represent the pore volume (cm^3/g) of these materials. (Reproduced from Ref. 1 with permission from The American Association for the Advancement of Science.)

highest surface area obtained by any porous materials. The size and shape of the pores or holes in these materials can be altered by merely modifying the structure of the linker or changing the metal salt. The main reason for this high surface area is the removal of the solvent from the pores without collapsing the pores using the supercritical CO_2 activation technique. To prepare MOFs with even higher surface area (ultrahigh porosity) requires an increase in storage space per weight of the material. Computational analysis of these kind of MOFs was revealed that the maximum surface area of MOF can be as high as $\sim$14600 m^2/g. Large space and higher surface area can be obtained using longer linkers which provides larger storage space and a greater number of adsorption sites. So the choice of organic linker and their positioning in the MOF structure are very important.

Pore size is another important factor that effect the gas separation performance of the MOFs, especially the gas selectivity and the flux of the gas through the pores. The distribution of pore sizes, pore shapes and pore volume of MOFs dictate the application of these MOFs. In general, the pores are classified into three different ranges as micropores (below 2 nm), mesopores (2 nm to 50 nm) and macropores (above 50 nm). For gas separation applications, pore diameter should be in the micropore range while having the high surface area. Most of the MOFs fall in the category of microporous solids.

Pore size is particularly important in the case of separation of gas mixtures based on their kinetic diameters. The kinetic diameter of different gases is presented in Figure 3.3, in comparison with the pore size of some of the zeolites.[13] MOFs can be synthesized with precise pore size in order to separate specific gas pair based on their kinetic diameters. MOF can also be post-functionalized in order to tune the pore size as needed for specific gas separation. The smaller gas can adsorb and pass through the pores of the MOFs while larger molecule rejected. The pores needs to be very defined and ordered in order to have high gas selectivity. Pore chemistry is also important. The composition of metal salts and linkers and their alignment dictates the pore chemistry. In the case of water and organic solvent separation, hydrophilic pore structures can adsorb water from organic solvents and hydrophobic pores adsorb organic solvent from water.[14] MOF pores can be tunable to control relative rates of adsorption and fluxes through the pores.

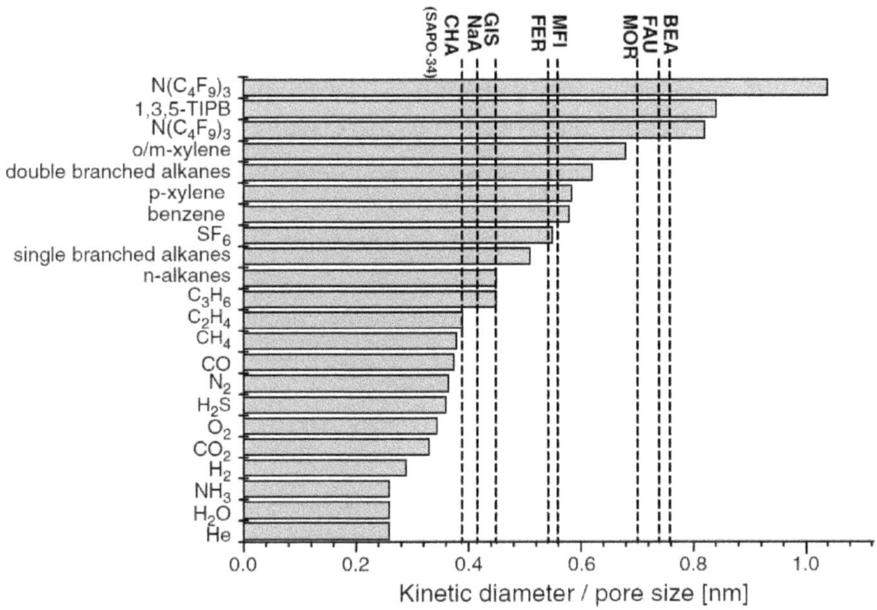

Figure 3.3. Comparison between the effective pore sizes of different zeolites and the kinetic diameters of gas molecules featured in reports on zeolite-based membrane reactor applications. (Reprinted from Ref. 13 with permission from Elsevier.)

3.2.2. *Chemical, thermal and mechanical stability of MOFs*

In the early stage of MOF research, the chemical, thermal, hydrothermal and mechanical stabilities of MOFs compared to other porous materials was a problem to apply them to the industry. It was a challenge to make chemically stable MOFs because of their susceptibility to linker or metal displacement reactions when treated with solvents over extended periods of time due to their weak coordination bonds.[15] Because of their easy structural tunability and easy functionalization, several chemically and thermally stable MOF has been synthesized. It is very important to study the thermal and chemical stability of the MOFs in order to find applications for them.

Some of the gas separation applications such as CO_2 capture from power plant flue gas, natural gas sweetening and synthesis gas purification, always contain moisture in them at elevated temperatures. The stability of some of the MOFs under humidity is not great. Hence, it is very important to study the hydrothermal stability of MOFs in order to apply them for the above mentioned gas separation applications. For example, ZIF-8, which was reported in 2006, is one of the first MOF that was extensively studied for its thermal and chemical stability.[16] The chemical stability of ZIF-8 and ZIF-11 was examined at typical industrial conditions. These materials were suspended in benzene, methanol, water, and different concentration of aqueous sodium hydroxide and boiled them at different times. PXRD patterns showed that both ZIF-8 and -11 maintained their full crystallinity even after 7 days of boiling them in organic solvents. Hydrothermal stability of ZIF-8 is also very good even in the presence of NaOH. High chemical and thermal stability of ZIF-8 was the results of having hydrophobic pore and surface of ZIF-8 and strong metal-organic bond strength.[16] The hydrothermal stability of several other MOFs was summarized in Figure 3.4.

MOFs were formed by coordination bonds between the metal salts and organic ligands. Coordination bonds are generally weak compared to the

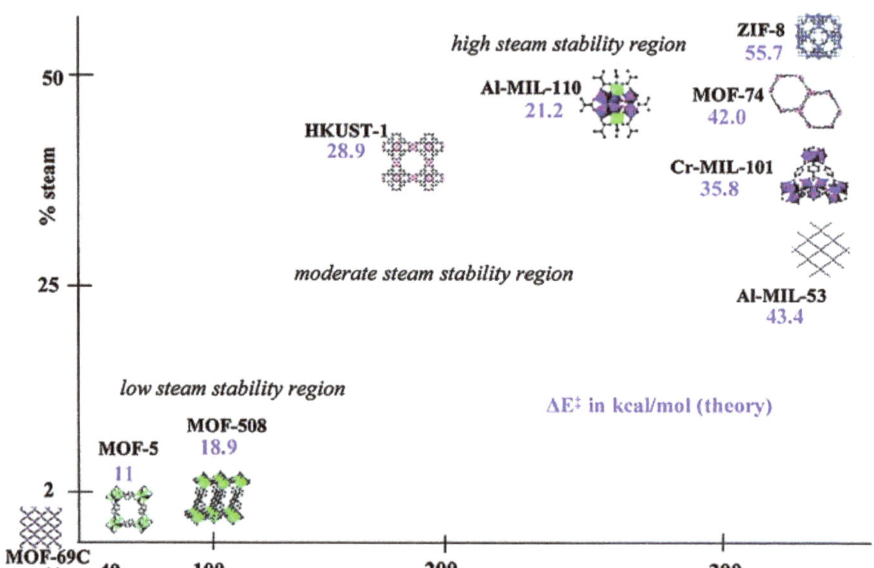

Figure 3.4. Steam stability map for several MOFs. The position of the structure for a given MOF represents its maximum structural stability by XRD measurement. The energy of activation for ligand displacement by a water molecule determined by molecular modeling is in kcal mol^{-1}. (Reprinted from Ref. 17 with permission from the American Chemical Society.)

covalent bond. The thermal stability of the MOFs is less compared to the other porous materials because of the breakages of the weak coordination bonds between metal ions and organic linkers. The thermal stability of the MOFs is rarely above 500°C and is low compared to the zeolites.[18] Thermal stability of the MOFs is generally related to metal node–organic linker bond strength and the number of linkers connected to each node. Thermal degradation of MOFs occur in several ways such as MOF amorphization, melting, node-cluster dehydration, or linker dehydrogenation or graphitization.[19] Thermal degradation of MOFs occur by starting with the breakage of coordination bond and then the decomposition of organic linker. Thermal stability of the MOFs can be measured using the thermal gravimetric analysis. It is very important to conduct the TGA experiments in different gas environments such as nitrogen and oxygen depending on the final application of the MOF. In some cases, TGA might provide misleading information.[20] As the temperature increases, the MOF might change the phase transitions which cannot be detected by TGA. Differential scanning calorimetry (DSC) can be used in order to probe these transitions. Also, in some cases, the partial collapse of the pores at higher temperatures cannot be detected by TGA. BET surface area analysis and powder XRD provides the partial collapse of the MOF crystal. So, thermal stability of the MOFs should be analyzed in combination with DSC, BET, and XRD. Yi *et al.* used powder XRD to probe the thermal and chemical stability of the modified UiO-67 MOFs as shown in Figure 3.5.[21] Use of rigid carboxylates such as benzene-1,3,5-tricarboxylate and benzene-1,4-dicarboxylate as the organic linker in the MOF synthesis yielded neutral, non-interpenetrated networks that decomposition temperatures in the range of 300–500°C.[3] According to Rosewell *et al.*, higher stability of these MOFs attributed to strong bond due to enhanced electrostatic attraction, and the size of the carboxylate functionality permits bridging of metal cations to produce rigid, geometrically defined clusters.[3]

Mechanical stability of the MOFs are important in order to apply them in industry in application like gas adsorption in packed bed and fluidized bed. The adsorbents, like MOFs, are usually made into industrially useful pellets and packed them in a column for gas separation applications. The elastic modulus of the MOFs are very low compared to other porous materials and other inorganic dense materials as shown in Figure 3.6.[22] The structure of

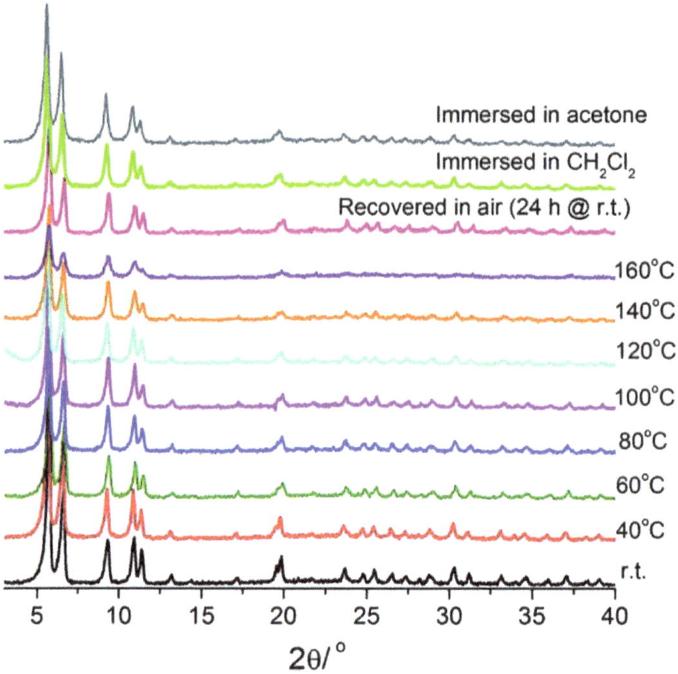

Figure 3.5. XRD patterns of UiO-67–Tz–NH_2 heated at various temperature, recovered in moist air, and immersed in different solvents. (Reproduced from Ref. 21 with permission from the Royal Society of Chemistry.)

the MOFs framework collapse or distort under shear stress or hydrostatic pressure due to the low elastic modulus.[23]

The mechanical properties of several zeolitic imidazolate frameworks (ZIFs) was studies using single-crystal nanoindentation studies and compared with several other inorganic solids.[24] It was observed that the elastic properties of ZIF crystal structures are strongly correlated to the rigidity and bulkiness of the substituted organic linkages. Wu *et al.* compared the mechanical properties of several well-known MOFs with the UiO-66-based MOFs as shown in Figure 3.7.[24] UiO-66 MOFs showed unusually high shear stability which is comparable to the zeolites and much higher than the other well-known MOFs such as MOF-5, ZIF-8 and HKUST-1. The shear modulus of UiO-66 is an order more than the other MOFs. The isoreticular series of UiO-type MOFs consists of a high degree of framework coordination between the dicarboxylate ligands and $Zr_6O_4(OH)_4$ nodes resulting in the high shear and bulk moduli compared to the other MOFs.[24] This excellent

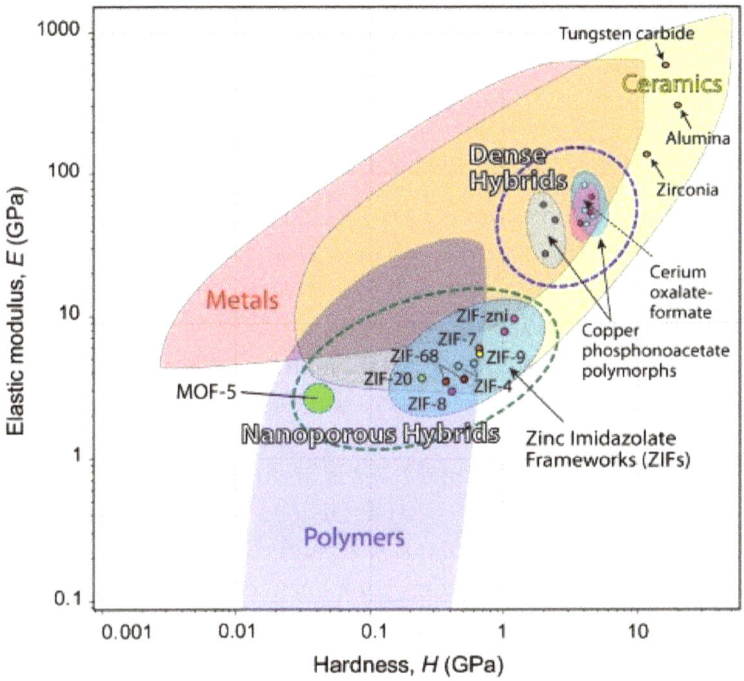

Figure 3.6. Elastic modulus versus hardness materials property map for dense and nanoporous hybrid framework materials shown alongside purely organic and inorganic materials. (Reproduced from Ref. 22 with permission from the Wiley.)

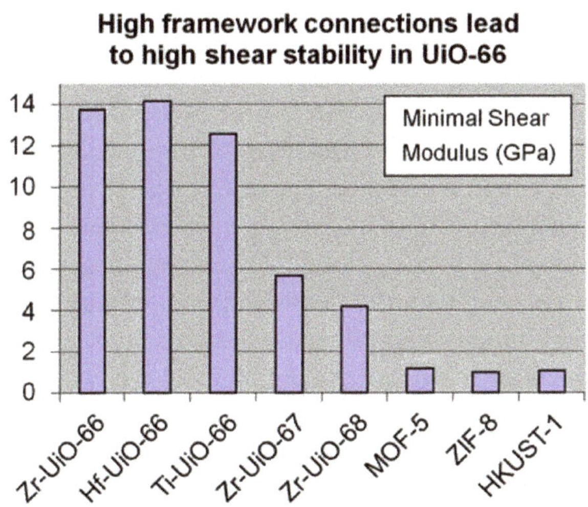

Figure 3.7. Comparison of mechanical properties of different metal organic frameworks. (Reprinted from Ref. 24 with permission from the American Chemical Society.)

mechanical properties along with chemical and thermal stabilities makes UiO-66-type MOFs outstanding candidates for practical application consideration.

3.3. Gas adsorption properties of the MOFs

Adsorption is a phenomenon of attraction of a molecule from a fluid phase to a solid surface, usually porous solids. Zeolites and carbon-based porous materials were used as adsorbents for a long time.[25,26] As a new generation of porous materials, MOFs shown promising properties for gas separation applications with large surface area per unit mass and organized pore structure.[27] The tunability of MOF structures and chemistry is the main advantage compared to the other porous materials. In this section, the gas separation using adsorption of gases on the MOFs is discussed.

The gas separation in MOFs can be divided as thermodynamic separation, kinetic separation, molecular sieving and conformational separation.[17] Preferential adsorption of a gaseous component due to the strong adsorbate-adsorbent concentrations compared to the other gas results in the gas separation and is called thermodynamic equilibrium effect. Another mechanism of gas separation is called the kinetic effect, in which one of the gaseous species diffuses in the pore faster than the other. Gas separation also occurs by molecular sieving in which one of the component enters the pores of the materials while the other is prevented from entering the pore. In another case, some light gases diffuses faster than the heavier gases and results in the separation of those gases called quantum sieving effect.

The adsorption-based gas separation in MOFs usually occurs by the adsorption one of the species by van der wall forces compared to the other species in the gas mixture. Another approach to achieve gas separations is through chemisorption of one of gaseous species selectively. Chemisorption usually takes place by reversible chemical reactions which involves the orbital overlap and charge transfer. Physisorption occurs by the electrostatic interactions and is much weaker than the chemisorption. Heat of adsorption of physisorption usually ranges between 5–50 KJ/mol while chemisorption is higher than the 50 KJ/mol. But in the case of gas separation application, the heat of reaction should be optimum in order to achieve high working capacity with minimum energy needed for adsorbent regeneration. As mentioned earlier, MOFs are highly tunable and can be modified and functionalized to achieve the physisorption and/or chemisorption properties. The gas selectivity in physisorption is also low compared to the chemisorption because of the low heat or energy challenging to target specific gas separation selectively. While in the case of chemisorption, selective towards the desired gas is very high compared to the other because of the strong binding energy.

Adsorption and desorption are the two main steps that needs to be optimized to become viable material for gas separation application. Adsorption process control the technical feasibility of a process and the desorption step controls its economic viability. The stronger the affinity between the adsorbate and adsorbent, the more difficult it is to desorb the gas. As the heat of sorption increases, the difficulty in desorbing the gas increases, resulting in higher energy requirement and results in the higher overall process cost. The desorption step has to be very carefully balanced with the adsorption step to design a successful adsorption process. Also, physical adsorbents are easy to regenerate compared to chemical adsorbents. Therefore, although physisorption is weaker than chemisorption, the efficiency of the gas separation process is high in the case of physical adsorbents.

Pressure swing adsorption (PSA) and temperature swing adsorption (TSA) are the two well-known processes for gas separation using adsorbents.[27] The gases are adsorbed at higher pressure and regenerate the adsorbent at lower pressures in the case of PSA. In the case of TSA, the gases are adsorbed at lower temperature and desorbs at higher temperature. The difference in the adsorption temperature and desorption temperature dictates the economics and energy requirements of the process. The equilibrium of adsorption establishes the thermodynamic limit of the adsorbent loading on the MOFs for a given fluid phase composition, temperature, and pressure. The time required to achieve the equilibrium state is also important in designing the adsorption process.

The strength of these interactions between the gases and MOFs depends on the gas dipole moment and the strength of the electric field produced by MOFs. The polarity and their polarization strengths are given in Figure 3.8.[28] Achieving high selectivity is almost impossible in the case of low and/or similar polarizable gases (e.g. CH_4/Kr or N_2/O_2) and is easy with the mixture of gases which have different polarizabilities. The other factor, MOF electric field strength, is determined by local spatial charge density. The magnitude of the electric field is determined by the local spatial

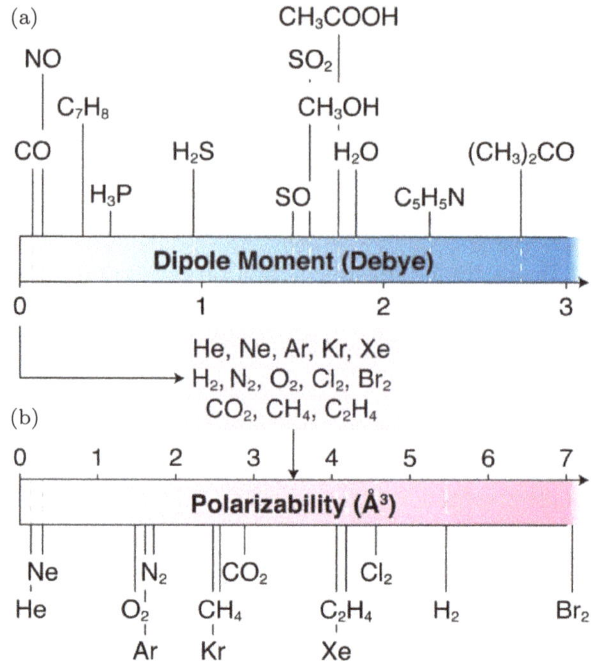

Figure 3.8. Permanent polarization (dipole moment, μ) of some familiar polar molecules (a) determines the interactions strength with the electric field produced by the framework. Most small gases feature no permanent dipole and their strength of interaction is determined by the magnitude of their polarizability (b). (Reprinted from Ref. 28 with permission from the American Chemical Society.)

charge density of the MOFs.[15] Open metal sites in some of the MOFs such as MOF-74, produce large electric fields and show high selectivity towards polar gases.[29] The organic linker of the MOF has been functionalized to increase the electric field strength in the pore centers. The MOF-74-based materials showed very high uptake of CO_2 because of the large quadrupole moment of CO_2, thus resulting in stronger interaction with the framework compared to other non-polar gases such as hydrogen and methane. There are several studies reported in the literature that helps in understanding the role of the metal ion in facilitating the adsorption properties of MOFs.[5]

Extensive research works have been done to make CO_2 adsorption favorable in MOFs and can be summarized into three main categories: catenation and interpenetration, chemical bonding enhancement, and electrostatic force involvement.[30] The CO_2 adsorption capacity of several MOFs are

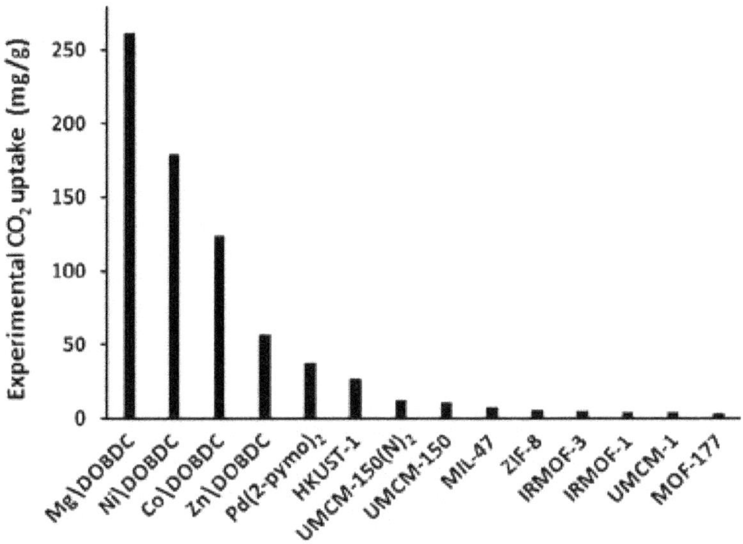

Figure 3.9. Experimental CO_2 uptake in different MOFs at 0.1 bar. Data were obtained at 293–298 K. (Reprinted from Ref. 31 with permission from the American Chemical Society.)

summarized in Figure 3.9.[31] The size of a pore is found to be critical to the adsorption affinity for light gases and a slight variation in pore size can cause a dramatic change in the adsorption affinity. Electrostatic force is generally introduced into MOF structures through metal ions doping and polar species modification. This modification introduces electrostatic force into the MOF structure and leads to considerable enhancement of the CO_2/N_2 selectivity. Amine functionalization of the surface and pore of the MOF is a common technique to improve the interaction between MOF and polar gases. As mentioned earlier in Chapter 2, the MOFs can be synthesized with amine functional group or amines can be introduced by post-synthetic modification.

MOFs can be categorized into rigid and flexible/dynamic classes depending on their structural behavior. Flexible MOFs possess dynamic pore structures that responds to pressure, temperature, and presence of guest molecules by breathing, swelling and linker rotation mechanisms while rigid MOFs have stable and robust pore structures.[32] The dynamic pore system of flexible MOF materials makes them excellent candidates for application in different kinds of separation processes especially gas separations using pressure swing adsorption process due to the differences in adsorption behavior of different gases. CID-3, a flexible MOF composed of [Zn(2,7-ndc)(bipy)]n layers was used to separate a gas mixture of CO_2, O_2 and N_2 using adsorption process and showed excellent gas separation performance of the material.[33] MIL-53 is another flexible MOF that has been studied extensively for gas separation applications. It was observed that MIL-53(Cr) structure was contracted with the adsorption of CO_2 on the hydroxyl groups in the framework.[34] There are several studies on the deformation mechanism of MIL-53 and its effect on gas separation mechanism.[35]

3.4. MOF membranes for gas separation applications

Gas separation membranes offer a number of benefits over other gas separation technologies. Conventional technologies such as the cryogenic distillation, amine absorption and adsorption on solids have several disadvantages like high energy requirement, chemical involvement, phase changes and complex equipment. Membrane process does not involve any phase transformation and the required membrane process equipment is simple with no moving parts, easy to operate and easy to scale up. Membranes have been widely used in various industrial separation applications for the last two decades. Currently, the industry is dominated by polymeric membranes that have been used in a wide variety of applications. Polymeric membranes such as cellulose acetate, polyimide and polyaramide can separate CO_2 from N_2 mixtures, but the high CO_2 pressures plasticize them and decrease their separation ability considerably.[36] Although the current membrane technology is dominated by polymers, these membranes suffer from low CO_2 permeability and selectivity, because they are based on a solution-diffusion mechanism, relying its separation ability mainly on the small size differences of the gas molecules.[37,38] Recently, research directed at the development and application of inorganic membranes is gaining momentum because of their excellent properties for molecular gas separations.[39] Despite much progress in the development of zeolite membranes, there is so far no industrial gas separation because of the fabrication issues like grain boundaries, costly support and reproducibility.

MOFs have many advantages such as tunable pore size, wide variety of chemical-physical properties and unlimited number of structures over

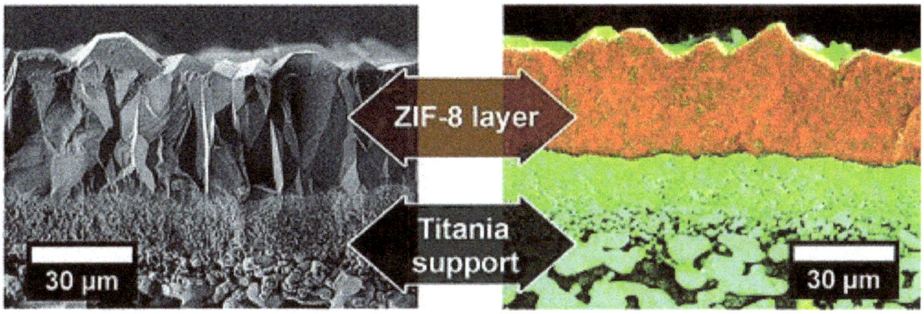

Figure 3.10. (left) SEM image of the cross section of a simply broken ZIF-8 membrane. (right) EDXS mapping of the sawn and polished ZIF-8 membrane (color code: orange, Zn; cyan, Ti). (Reprinted from Ref. 40 with permission from the American Chemical Society.)

zeolites and other inorganic materials as gas separation membranes. MOFs can be fabricated as a pure membrane on a porous support, an example of a MOF membranes (ZIF-8) is shown in Figure 3.10 which was fabricated on the Titania support by Caro and coworkers.[40] Although MOF membranes are very recently developed compared to zeolite membranes, they showed good gas separation performance and have high potential to be applicable in gas separations at industrial scale. Another unique feature of MOF membranes is related to its unprecedented high gas uptakes which in principle should translate into high gas selectivities. As compared to other microporous crystalline structures such as zeolites, MOFs (if prepared in membrane form) offer a wide range of pore sizes leading to infinite number of structures with flexible pore sizes and structures. Finally, due to the presence of organic linkers in MOF structures, proper functionalization of MOFs may result in pore size control as well as adsorption properties control. This in principle would allow one to rationally design tailored MOF membranes for gas separation applications.

MOF membranes have significant advantages over polymeric membranes. They do not deform under high CO_2 pressures, have high thermal, mechanical, and chemical stability, and good erosion resistance.[41] The molecular sieving ability of MOF membranes is well known and has been exploited for a vast number of relevant gas mixture separations.[42–45] Due to the processing difficulties with some of the MOFs, researchers found another method of membrane fabrication by mixing with the polymers, called mixed matrix membranes. The gas separation performance of the

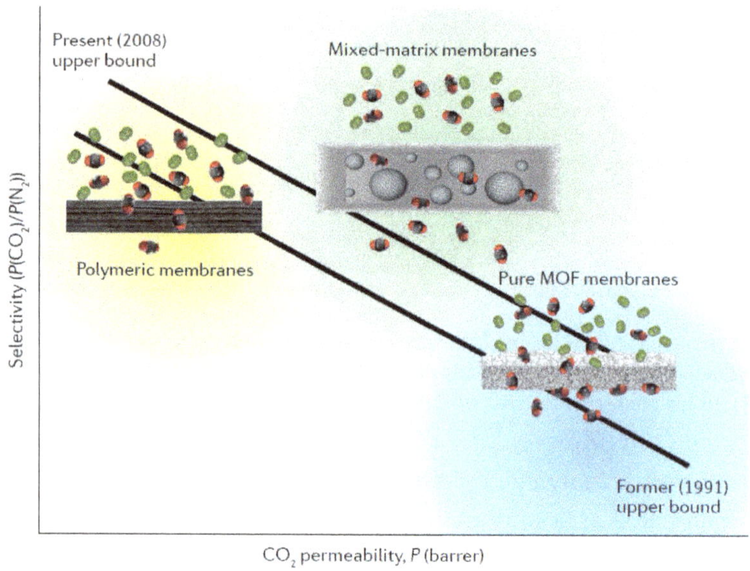

Figure 3.11. Selectivity versus CO_2 permeability compared to the Robeson upper bound curves for pure polymeric, mixed-matrix and pure MOF membranes. The relationship between the CO_2/N_2 gas pair selectivity and CO_2 permeability is compared to the Robeson upper bound curves for pure polymeric membranes, mixed-matrix membranes and pure metal-organic framework (MOF) membranes. The general trend is increased selectivity towards CO_2/N_2 with the use of mixed-matrix membranes (shown with MOF nanoparticles embedded within) and increased permeabilities for pure MOF membranes (shown on a porous substrate) when compared to traditional polymeric membranes. Each membrane depicts the separation of CO_2 from a gaseous mixture containing N_2. Color code for gas molecules: C, grey; O, red; and N, green. (Reproduced from Ref. 9 with permission from the Springer Nature.)

MOF membranes and MOF-based mixed matrix membranes is compared with the polymeric membranes as shown in Figure 3.11. Even though the above plot is made in the case of CO_2, this is true in the case of most other gas separation performance. As shown in Figure 3.11, the gas selectivity is still low compared to the pure polymer membranes mainly because of the fabrication issues which results in the defects in the membranes. Further research on the development of MOF membranes is needed to fabricate them reproducibly while having defect-free membranes with high gas selectivity. Even though MOF crystals exhibiting unique and remarkable separation properties, the same material may be not suitable for membrane fabrication due to limited adhesion of the membrane to the support, considerable adsorption induced stresses, poor crystal intergrowth, large crystal size leading to non-continuous membranes, poor reproducibility, etc. Cost of the MOF production should be minimized in order to use these MOFs for industrial applications.

References

1. Furukawa, H., Cordova, K. E., O'Keeffe, M. & Yaghi, O. M. The chemistry and applications of metal-organic frameworks. *Science* **341**, 1230444 (2013).
2. Li, J.-R., Kuppler, R. J. & Zhou, H.-C. Selective gas adsorption and separation in metal-organic frameworks. *Chem. Soc. Rev.* **38**, 1477 (2009).
3. Rowsell, J. L. C. & Yaghi, O. M. Metal-organic frameworks: A new class of porous materials. (2004). doi:10.1016/j.micromeso.2004.03.034
4. Li, J.-R., Sculley, J. & Zhou, H.-C. Metal-organic frameworks for separations. *Chem. Rev.* **112**, 869–932 (2012).
5. Sumida, K. *et al.* Carbon dioxide capture in metal-organic frameworks. *Chem. Rev.* **112**, 724–781 (2012).
6. Rowsell, J. L. C. & Yaghi, O. M. Metal-organic frameworks: A new class of porous materials. *Microporous Mesoporous Mater.* **73**, 3–14 (2004).
7. Davis, M. E. Ordered porous materials for emerging applications. *Nature* **417**, 813–821 (2002).
8. Venna, S. R. & Carreon, M. A. Metal-organic framework membranes for carbon dioxide separation. *Chem. Eng. Sci.* **124**, 3–19 (2015).
9. Trickett, C. A. *et al.* The chemistry of metal-organic frameworks for CO_2 capture, regeneration and conversion. *Nat. Rev. Mater.* **2**, 17045 (2017).
10. Yaghi, O. M. *et al.* Reticular synthesis and the design of new materials. *Nature* **423**, 705–714 (2003).
11. Farha, O. K. & Hupp, J. T. Rational Design, Synthesis, Purification, and activation of metal-organic framework materials. *Acc. Chem. Res.* **43**, 1166–1175 (2010).
12. Farha, O. K. *et al.* Metal-organic framework materials with ultrahigh surface areas: Is the sky the limit? *J. Am. Chem. Soc.* **134**, 15016–15021 (2012).
13. McLeary, E. E., Jansen, J. C. & Kapteijn, F. Zeolite based films, membranes and membrane reactors: Progress and prospects. *Microporous Mesoporous Mater.* **90**, 198–220 (2006).
14. Davis, M. E. Ordered porous materials for emerging applications. *Nature* **417**, 813–821 (2002).
15. Hendon, C. H., Rieth, A. J., Korzyński, M. D. & Dincă, M. Grand challenges and future opportunities for metal-organic frameworks. *ACS Cent. Sci.* **3**, 554–563 (2017).
16. Park, K. S. *et al.* Exceptional chemical and thermal stability of zeolitic imidazolate frameworks. *Proc. Natl. Acad. Sci. U. S. A.* **103**, 10186–10191 (2006).
17. Low, J. J., Benin, A. I., Jakubczak, P., Abrahamian, J. F., Faheem, S. A. & Willis, R. R. Virtual high throughput screening confirmed experimentally: Porous coordination polymer hydration. *J. Am. Chem. Soc.* **131**(43), 15834–15842 (2009).

18. Colombo, V. *et al.* High thermal and chemical stability in pyrazolate-bridged metal-organic frameworks with exposed metal sites. *Chem. Sci.* **2**, 1311 (2011).
19. Howarth, A. J. *et al.* Chemical, thermal and mechanical stabilities of metal-organic frameworks. *Nat. Rev. Mater.* **1**, 15018 (2016).
20. Howarth, A. J. *et al.* Best practices for the synthesis, activation, and characterization of metal-organic frameworks. *Chem. Mater.* **29**, 26–39 (2017).
21. Yi, X.-C., Xi, F.-G., Qi, Y. & Gao, E.-Q. Synthesis and click modification of an azido-functionalized Zr(iv) metal-organic framework and a catalytic study. *RSC Adv.* **5**, 893–900 (2015).
22. Allendorf, M. D., Schwartzberg, A., Stavila, V. & Talin, A. A. A Roadmap to implementing metal-organic frameworks in electronic devices: Challenges and critical directions. *Chem. — A Eur. J.* **17**, 11372–11388 (2011).
23. Hobday, C. L. *et al.* A Computational and experimental approach linking disorder, high-pressure behavior, and mechanical properties in UiO frameworks. *Angew. Chemie Int. Ed.* **55**, 2401–2405 (2016).
24. Wu, H., Yildirim, T. & Zhou, W. Exceptional mechanical stability of highly porous zirconium metal-organic framework UiO-66 and its important implications. *J. Phys. Chem. Lett.* **4**, 925–930 (2013).
25. Hauchhum, L. & Mahanta, P. Carbon dioxide adsorption on zeolites and activated carbon by pressure swing adsorption in a fixed bed. *Int. J. Energy Environ. Eng.* **5**, 349–356 (2014).
26. Chue, K. T., Kim, J. N., Yoo, Y. J., Cho, S. H. & Yang, R. T. Comparison of activated carbon and zeolite 13X for CO_2 recovery from flue gas by pressure swing adsorption. *Ind. Eng. Chem. Res.* **34**, 591–598 (1995).
27. Bae, Y.-S. & Snurr, R. Q. Development and evaluation of porous materials for carbon dioxide separation and capture. *Angew. Chemie Int. Ed.* **50**, 11586–11596 (2011).
28. Hendon, C. H., Rieth, A. J., Korzyński, M. D. & Dincă, M. Grand challenges and future opportunities for metal-organic frameworks. *ACS Cent. Sci.* **3**(6), 554–563 (2017).
29. McDonald, T. M. *et al.* Capture of carbon dioxide from air and flue gas in the alkylamine-appended metal-organic framework mmen-Mg_2 (dobpdc). *J. Am. Chem. Soc.* **134**, 7056–7065 (2012).
30. Liu, J., Thallapally, P. K., McGrail, B. P., Brown, D. R. & Liu, J. Progress in adsorption-based CO_2 capture by metal-organic frameworks. *Chem. Soc. Rev.* **41**, 2308–2322. (2012).
31. Yazaydin, A. O., Snurr, R. Q., Park, T. H., Koh K., Liu, J., Levan, M. D., Benin, A. I., Jakubczak, P., Lanuza, M., Galloway, D. B., Low, J. J. & Willis, R. R., Screening of metal-organic frameworks for carbon dioxide capture from flue gas using a combined experimental and modeling approach. *J. Am. Chem. Soc.* **131**(51), 18198–18199 (2009).
32. Schneemann, A. *et al.* Flexible metal-organic frameworks. *Chem. Soc. Rev.* **43**, 6062–6096 (2014).
33. Nakagawa, K. *et al.* Enhanced selectivity of CO_2 from a ternary gas mixture in an interdigitated porous framework. *Chem. Commun.* **46**, 4258 (2010).
34. Sandrine Bourrelly, *et al.* Different Adsorption Behaviors of Methane and Carbon Dioxide in the Isotypic Nanoporous Metal Terephthalates MIL-53 and MIL-47. (2005). doi:10.1021/JA054668V

35. Jiao, Y. *et al.* The studies on gas adsorption properties of MIL-53 series MOFs materials. *AIP Adv.* **7**, 85009 (2017).
36. Koros, W. J. & Mahajan, R. Pushing the limits on possibilities for large scale gas separation: Which strategies? *J. Memb. Sci.* **181**, 141 (2001).
37. Bredesen, R., Jordal, K. & Bolland, O. High-temperature membranes in power generation with CO_2 capture. *Chem. Eng. Process. Process Intensif.* **43**, 1129–1158 (2004).
38. Zolandz, R. R. & Fleming, G. K. Membrane Handbook, Chapman and Hall, New York. in *Ho, W.S.W.; Sirkar, K.K., Eds.;* (1992).
39. Caro, J. Hierarchy in inorganic membranes. *Chem. Soc. Rev.* **45**, 3468–3478 (2016).
40. Bux, H. *et al.* Zeolitic imidazolate framework membrane with molecular sieving properties by microwave-assisted solvothermal synthesis. *J. Am. Chem. Soc.* **131**, 16000–16001 (2009).
41. Kang, Z., Fan, L. & Sun, D. Recent advances and challenges of metal-organic framework membranes for gas separation. *J. Mater. Chem. A* **5**, 10073–10091 (2017).
42. Zhao, Z., Ma, X., Kasik, A., Li, Z. & Lin, Y. S. Gas separation properties of metal organic framework (MOF-5) membranes. *Ind. Eng. Chem. Res.* **52**, 1102–1108 (2013).
43. Caro, J. Are MOF membranes better in gas separation than those made of zeolites? *Curr. Opin. Chem. Eng.* **1**, 77–83 (2011).
44. Baker, R. W. & Low, B. T. Gas separation membrane materials: A perspective. *Macromolecules* **47**, 6999–7013 (2014).
45. Bohrman, J. A. & Carreon, M. A. Synthesis and CO_2/CH_4 separation performance of Bio-MOF-1 membranes. *Chem. Commun. (Camb).* **48**, 5130–5132 (2012).

Chapter 4

Membrane-based Gas Separations

4.1. Introduction

Membranes are thin semipermeable barriers that allows certain molecules to permeate through, while blocks other specific molecules from entering the membrane, thereby selectively separate a component from a mixture of components. This definition is necessarily broad because of the large variety of membrane materials separating an equally vast number of compounds in all phases. Membrane processes consist in the selective permeation of molecules through a thin layer of polymeric or inorganic material. Generally, any material that can be made into sufficiently thin and stable films can be used as membrane. The materials such as metal, glass, ceramics and polymers as well as ordered molecular monolayers of liquids can be used as membranes. Membrane has already been widely used in liquid separations such as micro-filtration, ultra-filtration, reverse osmosis, forward osmosis, desalination, gas separation and medical application. Development of gas separation membrane is one of most sought after research area over the last three decades. As explained in Chapter 3, membranes has several advantages compared to other competitive technologies such as distillation, absorption, etc.

Membrane gas separation is a pressure-driven process with different industrial applications in refineries and chemical industries. The membrane materials should be thermally and chemically robust, resistant to plasticization and aging affects to ensure continual performance over long time periods, as well as cost effective. Gas separation membranes were first commercialized in 1980 by Monsanto for hydrogen separation.[1] Extensive research resulted in innovations that improved the gas separation efficiency and membrane durability, making membrane gas separation commercially competitive with existing separation technologies. Gas separation membranes are now applied for the sweetening of natural gas by the removal of carbon dioxide, separation of nitrogen from air.

The two intrinsic criteria in determining the performance of a membrane are selectivity and permeability. Permeability and selectivity depends on the chemical nature and structure of the membranes. The economics of membrane processing also mainly determined by the selectivity and permeability of the membranes along with other costs. Low selectivity results in multi-step separation processes with higher operative complexity and costs. Smaller permeability causes an increase in the use of membrane modules or required membrane area and results in the low yield. A membrane with a suitable pore size capable of providing both high selectivity and permeability to a separation process is necessary for application. Currently, the most widely used membrane materials for gas separation are polymers because of the relatively low cost of the materials and easy processing, but they cannot operate at high temperatures or in harsh chemical environments.[2] Silica membranes are not stable in humid atmosphere at high temperatures, carbon membranes have low fluxes,[3] and metal membranes are vulnerable to surface poisoning and suffer from hydrogen embrittlement.[4] Zeolite membranes have the ability to sieve molecules at high temperature, high pressure, and in chemically harsh environments, and thus have significant potential.[5] But the tunability of the zeolite chemistry is very difficult and limited. It is very important to note that most of the zeolites were synthesized under aggressive conditions i.e. high temperature and pressure which are not viable for membrane fabrication at industrial scale.[6] MOF membranes can overcome the problems associated with zeolite membranes such as high temperature fabrication, high surface area and modification of pore chemistry and its size depending on the application.[7] After the invention of MOFs in the early 90's, MOF membrane research gained a lot of interests over the last two decades.

4.2. Definitions of gas separation

As mentioned earlier, the performance of a membrane process and its economics are largely determined by the membrane's transport properties like permeability and selectivity for a specific gas in a mixture. Permeability and selectivity are dictated by the gas diffusivity and adsorption properties of the

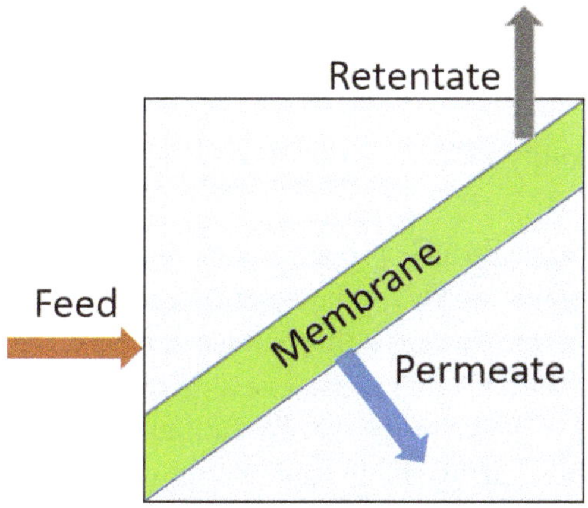

Figure 4.1. General principle of membrane gas separation. The gas passes on the surface of the membrane. Some of the components the gas mixture pass through the membrane while others are rejected by the membrane.

membrane material. A typical membrane process is showed in Figure 4.1. Permeate is the gas that transported across the membrane and retentate is the part of the gas that is rejected by the membrane. In this section, the detailed information about these parameters that are useful to design the membrane process are discussed.

4.2.1. *Permeability*

Fick's first law states that the molar flux due to diffusion is proportional to the concentration gradient.[8] Fick's first law of diffusion is mathematically expressed as shown in the following equation, where J is the flux through the membrane, D is the diffusivity, dc is the concentration difference of a gas across the membranes and dx is the thickness of the membrane.

$$J = -D\frac{dc}{dx}$$

The gases diffuse from high concertation region or partial pressure region to low concentration or low partial pressure region. Permeability is defined as the gas flux across unit membrane area under unit transmembrane driving force per unit membrane thickness. Permeability units are usually expressed in Barrer, named after Richard Barrer, where 1 Barrer $= 1 \times 10^{-10}$cm^3 (STP) cm/(cm^2 s cmHg).

$$\text{Permeability } (P) = \frac{\text{flux } (J)}{\text{pressure drop } (\Delta p)/\text{thickness } (l)}$$

Permeability is the intrinsic property of the material and is very useful to compare the gas separation performance of different materials. A membrane can separate gas mixture because different gases have different permeability through the membrane. In order to achieve high gas separation performance, a greater difference between gas permeabilities is preferred. Membranes with high permeability are useful in applications which involve large amount of gas processing such as power plant flue gas purification, natural gas sweetening and air separation.

The permeability (P) of a gas also defined as the product of diffusivity coefficient (D) and solubility coefficient (S) of a gas in the membrane material and expressed as $P = D.S$. The solubility coefficient of a gas is the ratio of the concentration of gas adsorbed or dissolved in the membrane at equilibrium to the gas pressure. Diffusivity is the mobility of the gas molecules in the membrane as they diffuse through the membranes. Solubility is a thermodynamic property and diffusivity is the kinetic property of the gases. In order achieve high permeabilities, the membrane should show either high diffusivity or high solubility or both. diffusivity usually depends on the kinetic diameter of gas molecules and solubility depends on the polarity and chemistry of the membrane materials.

The solubility coefficient of a gas depends on the interaction with the membrane material. Functional groups on the materials such as hydroxyl, amine interacts with polar gases such as CO_2, resulting in increased solubility and increase in the gas permeability. The solubility (S) depends on the temperature (T) and enthalpy of sorption (ΔH_s) and is described in terms of van't Hoff relationship.[9]

$$S = S_0 e^{\left(-\frac{\Delta H_S}{RT}\right)}$$

where, S_o is a constant, R is ideal gas constant and T is absolute temperature. Increase in the interaction between functional groups and polar penetrants molecules decreases ΔH_s, which lead to increase in gas solubility.

Similarly to solubility, diffusivity (D) is also expressed as;

$$D = D_0 e^{\left(-\frac{E_D}{RT}\right)}$$

where D_0 is the preexponential factor and E_D is the activation energy of diffusion. Diffusivity of the gases in the polymeric membranes depends on the penetrant size, fractional free volume, flexibility of the polymer chain flexibility, and polymer–gas interactions.[10] Similarly, the diffusion of gases in the porous membrane like MOFs depend on the pore size and its volume, kinetic diameter of gases, and the gas interactions with the material.

4.2.2. *Permeance*

The gas separation performance of a membrane is also reported in terms of permeance. Permeance is usually defined as the ratio of gas permeability to the thickness of the membranes. While permeability is the intrinsic property of the material, permeance is the property of the membrane structure and thickness of the membrane. In the case of asymmetric membranes, where the active membrane thickness is not precisely known, the flux is often reported in terms of permeance. Permeance is used more frequently in the membrane industry and is expressed in gas permeance units (GPU), where 1 GPU = 1×10^{-6}cm^3 (STP)/(cm^{2}s cmHg). The gas permeance of a membrane can be directly calculated for a membrane thickness by dividing the gas permeability with the thickness of the membrane in microns (Permeance = Permeability/membranes thickness(microns)). Most of the polymer literature and industries used GPU or cm^3 (STP)/cm^{2}s cmHg as the units of permeance. Inorganic membrane literature also used mol/(m^2. s. bar) in most cases to express the gas permeances and some of them also used GPU units. An article on units of gas permeability constants published in 1975 summarized the different units of gas permeability and their conversions.[11] Here are some of the conversions between different units of gas permeances.

$$1\,\text{GPU} = 10^{-6}\text{cm}^3(\text{STP})/(\text{cm}^2.\ \text{sec.}\ \text{cmHg})$$

$$\begin{aligned} 1\,\text{GPU} &= 3.35 \times 10^{-10}\text{mol}/(\text{m}^2.\ \text{s.}\ \text{Pa}) \\ &= 3.35 \times 10^{-5}\text{mol}/(\text{m}^2.\ \text{s.}\ \text{bar}) \end{aligned}$$

$$1\,\text{GPU} = 6.25 \times 10^{-4}\text{ft}^3(\text{STP})/(\text{ft}^2.\ \text{hr.}\ \text{psi})$$
$$1\ \text{Barrer} = 10^{-10}\text{cm}^3(\text{STP}).\ \text{cm}/(\text{cm}^2.\ \text{sec.}\ \text{cmHg})$$
$$= 1\,\text{GPU} \times (\text{thickness in microns})$$

Permeability is very useful to compare the properties of different materials and useful in the membrane materials research and development stage. But industries preferred the use of gas permeance as it is the true measure of the membrane module and is very useful for process design. Gas permeance dictates the required membrane area to perform a gas separation and higher permeances reduces the required membrane area and reduced size of the membrane module.[12] One way to increase the gas permeance is to reduce the membrane thickness. As we lower the thickness of the membranes, the permeance of the membrane increases significantly. But at certain optimum thickness of the membrane, the permeance won't increase because of the resistance in the support on which the membrane was fabricated. In order to further increase the permeance, the support needs to be modified to obtain high gas fluxes and minimize the resistance to gases. Another method is to increase the intrinsic permeability of the membrane material by modifying the membrane chemistry itself in such a way that increases the diffusion of gases in the membrane.[13,14]

4.2.3. *Selectivity*

Another parameter that is very useful to characterize the efficiency of a membrane is the selectivity. The ideal selectivity of a membrane is the ratio between the permeability (or permeance) of gas $A(P_A)$ and that of gas $B(P_B)$.

$$\text{ideal selectivity } (\alpha_{A/B}) = \frac{P_A}{P_B}$$

The ideal gas selectivity is an intrinsic property of a membrane material and the particular gas pair at particular temperature and pressure. Most of the literature reports present the ideal selectivity values, but mixed gas selectivity measurements are important in order to study the applicability of the materials in the industry. Similar to permeance, separation factor is an important parameter to design a membrane separation process and to determine efficiency of the membrane process. The separation factor is not only a material property, but also depends on the process conditions such as the driving force.[15] The separation factor of component A over component B is defined as the ratio of concentration of component A ($x_{A,p}$) and component B ($x_{B,p}$) in permeate side to the concentration of component A ($x_{A,f}$) and component $B(x_{B,f})$ in the feed stream.

$$\text{separation factor} = \frac{x_{A,p}/x_{B,p}}{x_{A,f}/x_{B,f}}$$

In the practical gas separation processes by industries, mixed gases are used instead of pure gases. The ideal selectivity and mixed gas selectivity might be similar if the gases in a mixture do not interact strongly with the membrane material, for example nitrogen and oxygen. But in the case of gas mixtures which contain polar and/or condensable gases, the ideal and mixed gas selectivities are very different because of the interaction of these gases with the membrane. With high selectivity and high permeability membranes, the separation of gases can be achieved in a single step at the optimum pressure ratio conditions which reduces the complexity and cost of the process.

The separation ability of a membrane, $\alpha_{A/Bj}$ is also defined in terms of ratios of diffusivity coefficient (D) and sorption coefficient (S) as shown below. α_S is the solubility selectivity and α_D is the diffusivity selectivity.

$$\alpha_{A/B} = \frac{P_A}{P_B} = \left(\frac{S_A}{S_B}\right) \times \left(\frac{D_A}{D_B}\right) = \alpha_s \times \alpha_D$$

The selectivity and permeability of membranes are most easily visualized using a Robeson plot.[16,17] Robeson plotted the selectivity vs permeability of different gases in several polymeric membranes. An example of Robeson upper bound plot for CO_2/N_2 gas separation is shown in Figure 4.2.[17] It was observed that selectivity decreases with increasing gas permeability of the more permeable gas component and resulting trade-off relationships yields a upper bound of the gas separation performance. Although this plot made for the polymeric membranes, MOF membrane researchers used this plot in order to compare the gas separation performance of MOF membranes with the polymeric membranes.

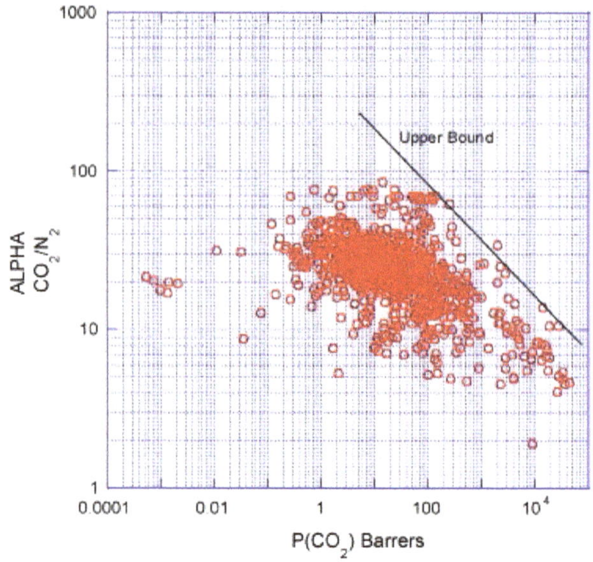

Figure 4.2. Upper bound correlation for CO_2/N_2 separation. (Reprinted from Ref. 17 with permission from Elsevier.)

4.3. Gas separation mechanism in MOF membranes

In dense polymer membranes, the gas separation occurs based on solution-diffusion mechanism. According to the solution–diffusion mechanism, the selectivity is determined by two factors: first, the components solubility which depends on the interaction of the polymer matrix and the component to be separated. Second, the diffusivity, which is mainly governed by the diameter of the components to be separated but also by the mobility of the polymer chains. But in the case of porous membranes the gas separation mechanism is much different than the polymeric membranes. There are three main transport mechanisms by which gas separation occurs in inorganic porous membranes such as metal organic frameworks, zeolites, and porous carbon. The basis of these transport mechanism, as shown in Figure 4.3, are the molecular weight (Knudsen diffusion), the size of molecules (molecular sieving) to be separated, and surface interactions (surface diffusion). The type of gas transport mechanism depends on the properties of the gas molecules and microstructure of the porous material. The details of the above mentioned three gas transport mechanisms in the porous membranes (as shown in Figure 4.3) are provided in this section.

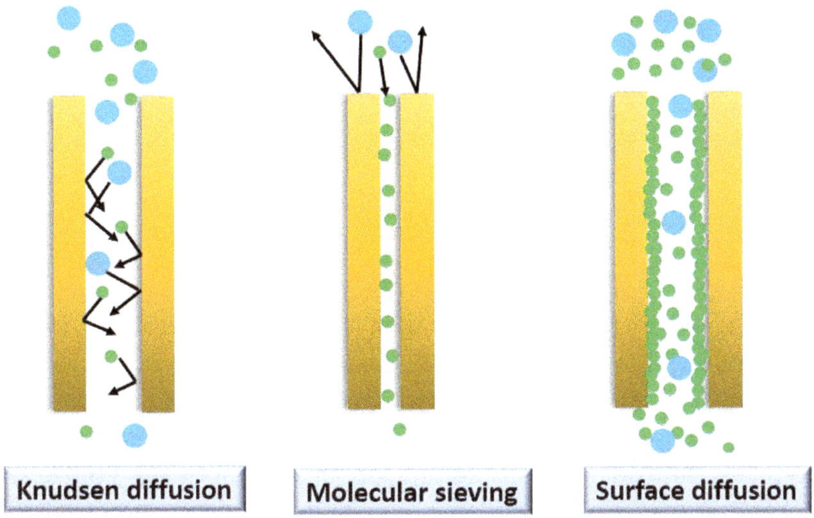

Figure 4.3. Schematic representation of the gas separation mechanisms in porous membranes by Knudsen diffusion, Molecular sieving and surface diffusion.

4.3.1. *Knudsen diffusion*

Knudsen diffusion occurs in the gas phase through the pores in the membrane layer whose pore diameters are smaller than the mean free path dimensions of the molecules in the gas mixture.[18] Knudsen number is typically used to describe the Knudsen transport mechanism of gases.[19,20] Knudsen number (K_n) is the ratio of the gas mean free path to the pore size of the materials ($K_n = \lambda/d_p$), where d_p is the diameter of the pores, and λ is the gas mean free path. The mean free path of the gas molecules (defined as the average distance a molecule travels between two collisions) is given by;

$$\lambda = \frac{1}{\sqrt{2}\,\pi\, d_A^2 \left(\frac{p_A\, N}{R\, T}\right)}$$

where d_A is the effective diameter of the molecule, and N is Avogadro's Number, R is the gas constant, T is temperature, p_A is the partial pressure.

Different diffusion regimes based on the Knudsen number (K_n) is given in Table 4.1.[21] If λ is greater than 10 times of the pore diameter, the movement of molecules inside the narrow pore channels takes place through

Table 4.1. Different diffusion regimes based on the Knudsen number.[21]

Diffusion regime	Mean Free Path relative to Pore Size
Knudsen	$\lambda > 10\,d_p$
Bulk	$\lambda < 0.1\,d_p$
Transition	$10\,d_p > \lambda > 0.1\,d_p$

collision of the diffusing molecules with the surface rather than with each other and the gas transport occurs by Knudsen diffusion. If λ is less than the 0.1 times of pore diameter, collisions and interactions between gas molecules themselves become dominant compared to molecular diffusion and viscous diffusion. In transition region, the gas molecules collide with each other as well as the pore walls, results in the contribution from both the Knudsen diffusion and bulk diffusion. The gas diffusion coefficients in solution-diffusion mechanism of polymeric membranes are typically much lower than the Knudsen, transition or bulk transport mechanism. Different molecular mass gas molecules have different interaction energy with the pore walls. In this regime, the presence of other gases no longer affects the transport, and the flux depends only on the density gradient of the gases. The relative diffusion rate of each component in Knudsen regime is inversely proportional to the square root of its molecular weight as shown in the equation below.

$$\frac{D_i}{D_j} \propto \sqrt{\frac{MW_i}{MW_j}}$$

where, D_i and D_j are the diffusive coefficient of i and j, MW_i and MW_j are the molecular weight of component i and j.

According to Knudsen diffusion, N_2 molecules preferentially permeate in the case of CO_2/N_2 separation. The selectivity of CO_2 with respect to N_2, CH_4, and H_2 by Knudsen diffusion will be 0.8, 0.6, and 0.2, respectively. Hence, the selectivity of CO_2 achievable by the Knudsen mechanism is very low and not attractive in this particular gas mixture. Several other gas mixture Knudsen selectivity values are provided in the Table 4.2. In Knudsen diffusion lighter gases have higher diffusivities than heavier gases. The flux

Table 4.2. Selectivity of some of the gas mixtures by Knudsen diffusion.

Gas mixture	Knudsen selectivity
CO_2/N_2	0.80
O_2/N_2	0.94
CO_2/O_2	0.94
CO_2/H_2	0.21
CO_2/CH_4	0.60
CO_2/H_2S	0.88
CO_2/CO	0.80
CO_2/H_2O	0.64
H_2/He	1.41
H_2/N_2	3.74
N_2/CH_4	0.76
C_3H_8/C_3H_6	0.98
C_2H_6/C_2H_4	0.97

of the gases through the membrane by Knudsen diffusion can be calculated as follows.

$$\text{Knudsen diffusion coefficient} = D_k = \frac{du}{3\tau}$$

where, d = diameter of the pore, u = average gas velocity, τ = tortuosity factor ($\tau = 1$, assuming the pores of MOFs are cylindrical and straight).

$$u = \text{average molecular speed} = \sqrt{\frac{8RT}{\pi m}}$$

where, R = universal gas constant, T = temperature, m = molecular weight of the gas.

Therefore, D_k can be expressed as:

$$D_k = \frac{d}{3}\sqrt{\frac{8RT}{\pi m}}$$

Flux in Knudsen regime of component i is given by,

$$J_{ik} = \frac{n\pi d^2 P D_k}{4RTL}$$

Substituting D_k in the above equation, the flux in Knudsen regime becomes,

$$J_{ik} = \frac{nd^3P}{6L}\sqrt{\frac{2\pi}{RTm_i}}$$

where, n is the surface concentration of pores, L is the thickness of the membrane, P is the log mean pressure drop across the membrane.

4.3.2. *Molecular sieving*

The most widely known gas separation mechanism in the case of porous materials such as MOFs is molecular sieving or size sieving.[22] As mentioned in Chapter 3, the MOFs have very defined pores and the pore size in MOFs can be tunable by synthesis modification or post-synthesis modification.

The pore size of the MOF membranes is between the smaller gas molecule and larger gas molecule as depicted in Figure 4.3. Kinetic diameters of common gases are given in Figure 3.3 and Table 4.3. According to the molecular sieving mechanism, the separation is caused by the passage of smaller molecules of a gas mixture through the pores of the MOF membranes, while the larger molecules cannot enter into these pores. But a very fine-tuning of the membrane pore sizes is required to achieve the desired

Table 4.3. Kinetic diameter and molecular weight of some of the common gases. (Reproduced from Ref. 23 with permission from Wiley.)

Gas	Kinetic diameter (nm)
Carbon dioxide	0.330
Nitrogen	0.364
Oxygen	0.346
Hydrogen	0.289
Helium	0.260
Methane	0.380
Propane	0.430
Propylene	0.450
Ethane	0.444
Ethylene	0.390
Hydrogen sulfide	0.360
Water	0.265
Argon	0.340
CO	0.376

separation efficiency especially in the case similar to size gas molecules. For example, the separation of O_2/N_2 by molecular sieving is very difficult because of their very similar sizes and pores of the material should be very defined and strong. Molecular sieving is very useful in the case of very different size molecules, such as H_2 and CO_2, H_2 and hydrocarbons, etc. MOF pores, in some cases, are flexible and difficult to separate the gases by molecular sieving with these particular MOFs such as MIL-53. But, metal-organic framework chemistry is developed to the point that the pore size can be tunable very precisely depending on the application. In the case of other porous materials such as zeolites, porous carbon cases, the pores are very rigid, but the tunability of the pore sizes is a big hurdle with these materials.

4.3.3. *Surface diffusion*

In the surface diffusion mechanism, the diffusing species adsorb on the walls of the pore, and then readily transport along the pore surface in the direction of decreasing surface concentration or partial pressure.[24] In general terms, surface diffusion is the adsorption of a component on a site and jump from one site to another through strong interaction as shown in Figure 4.3. Typically, the molecules with larger molecular weight and with larger polarity and polarizability are selectively adsorbed on the membrane surface.[25] Surface diffusion is generally used to separate polar and condensable gases such as carbon dioxide, propane, ethane from non-polar and non-condensable gases such as nitrogen, oxygen, etc. The diffusivity in surface diffusion mechanism are affected by the strength of the surface-gas interaction, orientation of the surface lattice, attraction and repulsion between surface species and chemical potential gradients.[25] The adsorbed species on the membrane pores can also drastically reduce the transport of other species across the pore by reducing the size of accessible void space through the pore. This effect introduces non-adsorptive separation selectivity for the adsorbed species when the pore size is between two to three molecular diameters of the adsorbed species. The interaction between a gas and the pore can also be introduced by modifying the adsorbent layers.

Multilayer diffusion occurs when species adsorb in several layers. A gas mixture permeates through the pores of membranes; vapor species condense in the pores at a pressure lower than the saturation pressure at a given temperature. The multilayer diffusional flux is much larger than the gas phase flux.

In some cases, the pore of the membrane filled with the condensable gas in multilayers and transport across the membranes, is called the capillary condensation mechanism. The important aspect of the capillary condensation is that the pore is blocked by condensate, preventing gas transport of other components of the gas mixture. These aspects can result in increased gas selectivity of the polar and/or condensable gas. The condensation pressure depends on the pore size and shape and also the strength of the interaction between the fluid and pore walls.

4.4. Gas separation performance measurement techniques

The pure and mixed gas separation performance measurement of the membranes using different techniques was summarized in a great way by Haiqing Lin and Benny Freeman in Section 7.6 of *Handbook of Materials Measurement Methods*.[26] Brief overview of the gas separation performance measurement techniques are presented in this section. Pure gas and mixed gas permeation is usually analyzed using constant volume-variable pressure (isochoric gas permeation system) or a constant pressure-variable volume (isobaric gas permeation system) equipment. Constant pressure-variable volume system is suitable for membranes that can have the permeate gas flow rate greater than 0.3 cm^3/min which can be detected by the bubble flow meter.[26] Lower permeate flow rates can be measured using highly advanced and accurate mass flow meters. Constant volume-variable pressure permeation system can be used for a wide range of membrane permeabilities by having appropriate downstream volume. The detailed working principles of these gas separation performance testing systems are provided here in the following sections.

4.4.1. *Constant volume–variable pressure method*

An example of constant volume–variable pressure (isochoric) gas permeation testing system is presented in Figure 4.4. In a typical experiment, the membrane was loaded into a membrane module. The upstream and downstream of the membrane usually de-gassed using a vacuum pump for several hours in order to remove the adsorbed gases from the membranes. The leak rate into the system should be measured by isolating the permeation system from the vacuum pump. In order to measure permeability accurately, the

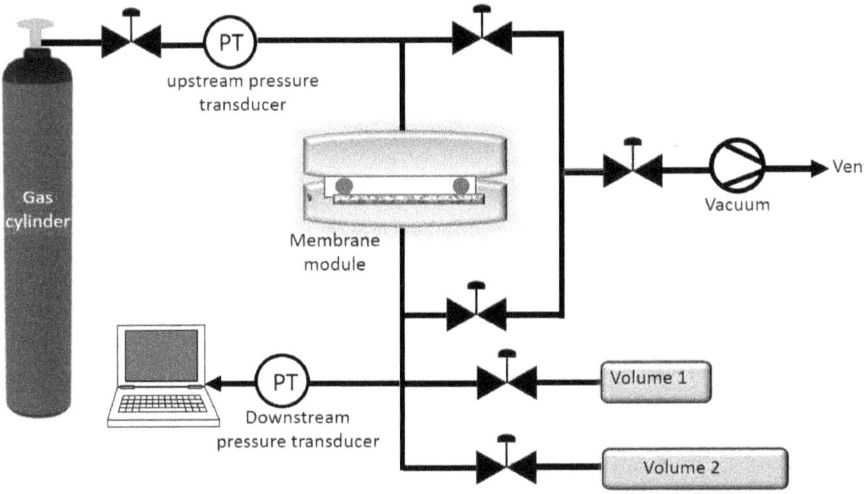

Figure 4.4. A typical constant-volume variable-pressure gas permeation performance system.

leak rate should be one to two orders lower than the steady state pressure rise.

The feed gas needs to be introduced to the upstream side of the membrane at the desired pressure. The upstream pressure should be much higher than the downstream pressure to maintain an effectively constant pressure difference across the membrane. The gases permeate onto the other side of the membrane due to the pressure driving force. The permeability of the membrane can be analyzed by monitoring the rate of pressure rise in the downstream volume as a function of time. Accurate measurement of downstream pressure rise as a function of time is very important and usually measured by using a commercially available pressure transducers such as Baratron 627D capacitance manometer (MKS, Wilmington, MA).[11] The maximum pressure that can be readable using this pressure transducer usually is about 10 torr and other options are available for higher pressure reading. The pressure transducer can be connected to the computer and record the real time pressure rise in the downstream using the software like Labview. In order to calculate the gas permeability properties of the membrane accurately, measurement of the downstream volume without error is important. Measurement techniques for the downstream volume is

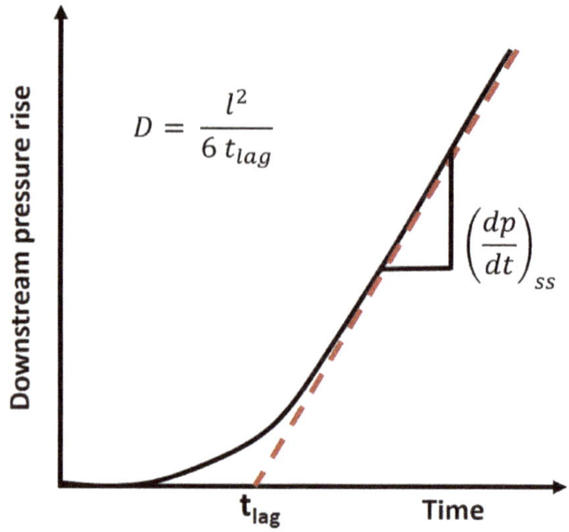

Figure 4.5. Typical pressure rise rate in the constant volume–variable pressure system.

provided in detail in the literature.[26] A typical pressure rise plot is shown in Figure 4.5. Once the steady state pressure rise is achieved, permeability of the membrane can be calculated by using the following equation. The ideal selectivity of the membrane is the ratio of the permeabilities of the individual gases.

$$P = l\, \frac{V_d}{p_u.A.R.T}\left[\left(\frac{dp_d}{dt}\right)_{ss} - \left(\frac{dp_d}{dt}\right)_{leak}\right]$$

where, V_d = downstream volume (cm^3), l = film thickness (cm), p_2 = upstream absolute pressure (cmHg), A = area of the membrane (cm^2), T = Temperature (K), R = gas constant (cm^{3}cmHg/mol$K$), $(dp_1/dt)_{ss}$ = rate of downstream pressure rise during testing (cmHg/sec), $(dp_1/dt)_{leak}$ = rate of downstream pressure rise under vacuum (cmHg/sec).

Due to the ability to measure pressure accurately, the constant–volume variable–pressure system is able to measure a wide range of flux values, including materials exhibiting a wide range of flux values. But in some cases, steady-state pressure rise cannot be achieved before the downstream pressure increases to the maximum limit of the downstream pressure transducer. In order to avoid this situation, the system can be designed with two downstream volumes of different sizes. Depending on the flux of the membrane, one volume or two volumes can be used to measure the pressure rise rate.

Constant volume–variable pressure system is also very useful in calculating the diffusivity and solubility of the membranes. Using this instrument, permeability, diffusivity and solubility can be estimated from a single experiment. The intercept of the steady-state region with the time axis is defined as the time lag (t_{lag}) as shown in Figure 4.5. The diffusion coefficient is given by: $D = l^2/6t_{lag}$ where l is the thickness of the membrane. From measurements of permeability and diffusivity, solubility can be estimated according to: $S = P/D$.

4.4.2. *Constant pressure–variable volume method*

Constant pressure-variable volume permeation (isobaric) system is shown in Figure 4.6. This system is very useful for higher flux membranes in

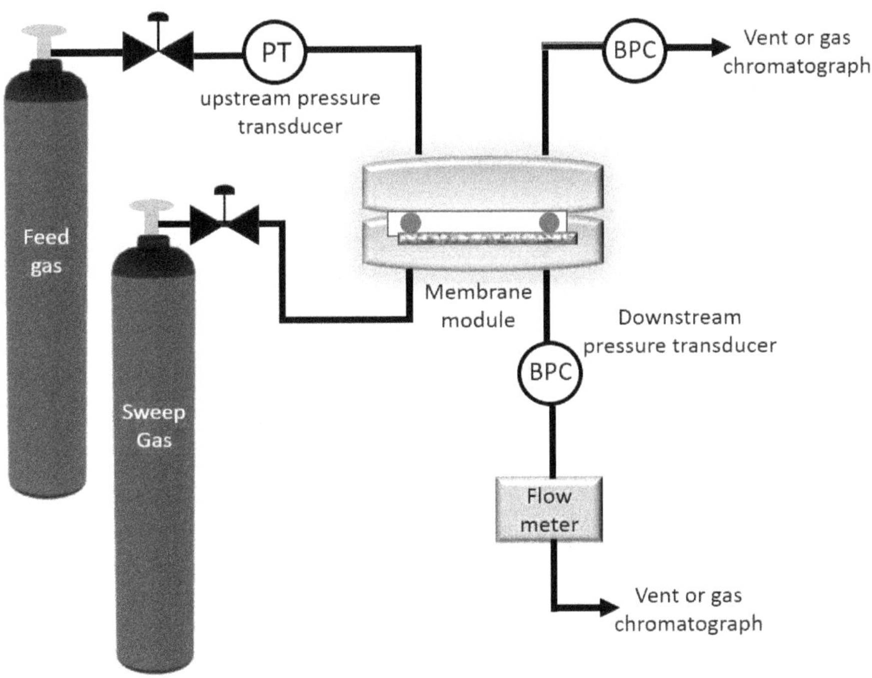

Figure 4.6. A typical constant–pressure variable–volume gas permeation performance system.

which the measurement of permeability is not possible with other system because of the faster pressure rise. Both the permeate side and retentate side are purged with an inert gas to remove any impurities from the gas lines before introducing the feed gas. The upstream and downstream pressure of the system can be controlled by back pressure controllers or needle valves. The feed gas enters the membrane module at a desired pressure and leaves as a residue stream. The gas which was diffused through the membrane is removed by using a sweep gas typically helium or argon. Measurement of downstream flow rate is very important to calculate the steady state gas flux which in turn used to calculate the permeability or permenace of the membrane. Permeate flow rate is measured typically using soap bubble flow meter. In some of the newer systems, mass flow meter can be used. The accuracy of this method depends on the sensitivity of the bubble flow meter or mass flow controller to measure the flow rate. Typically, the accuracy of the bubble flow meter becomes higher as the capillary tube diameter decreases.[26] After the introduction of the feed gas, the system needs to get steady state before collecting data. Permeability is calculated by using the following equation.

$$P = l\,\frac{273.\,p_{atm}}{T\,.\,\Delta p.\,76}\,\frac{dV}{A.\,dt}$$

where, l is the thickness of the membrane, T is temperature of measurement, Δp is the pressure drop across the membrane, A is the area of the membrane and dv/dt is the permeate flow rate.

4.4.3. *Mixed gas performance testing*

Gas separation performance testing of membranes using mixed gases is most relevant and useful for industrial applications. Mixed gas testing also provides the real gas separation performance of membranes and long term stability of the membranes in the presence of mixed gases. The permeability and selectivity of the membranes under mixed gases can be measured using constant volume–variable pressure or constant pressure–variable volume techniques. But the permeate stream needs to connect to gas composition measurement instrument, typically gas chromatograph, to analyze the components that permeated through the membrane. In the case of mixed gases, the gases mixed using mass flow controllers or using premixed gas cylinders. The mixed gases can be composed of several gases and number of gases is limited by the gas chromatograph which should have the ability to measure the fraction of each individual gases in the permeate stream. The composition of gases in the permeate can be calculated using the GC and permeate flow rate. Typically, a sweep gas is used to transport the permeated gases to the gas chromatograph. Sweep gas can be the same as the carrier gas of the GC, typically helium and argon. Carrier gas thermal conductivity should be different from the gases of interest in order to accurately measure the composition of gases using thermal conductivity detector. In the case of constant volume–variable pressure method, the gas collected in the downstream volume is transported to the GC once the steady state is reached. In the case of mixed gases, well mixing of the gas near the surface of the membrane is very important in order to avoid the concentration polarization effects. A specially designed cell can be used to ensure the well mixing of the feed as shown in other reports.[26] The stage cut, ratio of the permeate flow rate to feed flow rate, should be less than 1% to ensure that there is no concentration polarization near the membrane surface.

4.5. Gas flow configurations and membrane module types

The membranes, polymeric and inorganic membrane, are usually fabricated in two forms; flat sheet and tubular. A very small area of the high performance membrane is needed to test the gas separation performance of the materials at laboratory scale. However, a very large area membrane must be assembled as large modules for industrial applications. Membrane module, building block of a membrane system, is a unit composed of membrane with feed and permeate stream lines. Flat sheet and tubular membranes essentially can be packed into the modules in three different ways; (1) plate and frame, (2) spiral wound and (3) tubular. Flat sheet membranes are packed into plate and frame and spiral wound modules. Cylindrical hollow membranes such as tubular, capillary and hollow fiber are packed in tubular fashion. The main difference between the different tubular membranes is the diameter of the tube. The selection of a module type for a specific application depends on several factors. For gas separation applications, the module cost is an important factor which depends on the vessel in which the membranes are packed and the way of packing, which in turn depends on the pressure and temperature of the application.

4.5.1. *Plate and frame*

A typical design of the plate and frame module is shown in Figure 4.7.[27] The most important elements of the module construction are flat membrane, supporting spacer, and feed distribution plate. The plate and frame module design provides a configuration which is closest to the flat membranes used in the laboratory. Plate and Frame membrane systems utilize membranes laid on top of support plates that channel the permeate, which in turn is held together by a frame-like support. In a plate and frame module configuration, sets of two membranes are placed in a sandwich-like fashion with their feed sides facing each other. In each feed and permeate compartment, a suitable spacer is placed. Flat sheets used in plate-and-frame modules can be circular, square, or rectangle in cross-section. A plate-and-frame stack is build up by the number of membrane sets needed for a given membrane area equipped with sealing rings and two end plates. The packing density

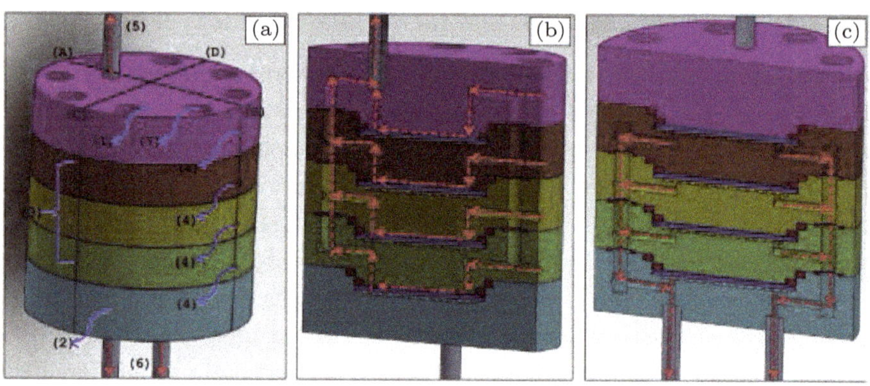

1. Top-plate	3. Body-plate	5. Retentate side tube	7. Screws holes
2. Bottom-plate	4. Feed inlet hole	6. Permeate side tube	

Figure 4.7. Schematic of plate-and-frame type membrane module for easy numbering-up: (a) is the assembled module; (b) is the cross-sectional view cut along (A) to (B); (c) is the cross-sectional view cut along (C) to (D); the dotted line in (b) is the flow line of the feed gas; the dotted line in (c) is the flow line of the permeated gas. (Reprinted from Ref. 27 with permission from Elsevier.)

(membrane surface per module volume) of such modules is low and about 100–400 m^2/m^3.

4.5.2. *Spiral wound*

Flat sheets can also be fabricated into spiral-wound modules. It provides large specific surface area at relatively low costs compared to the plate and frame modules. In a spiral wound module, two membrane sheets are separated by a spacer sheet to form a channel for flow of feed or permeate.[28] Feed gas flows through feed channel in axial direction and the gas permeates through the several layer of the membranes in radial direction and enter into the central perforated pipe as shown in Figure 4.8.[28] The packing density of spiral-wound modules is higher than the packing density of plate and frame modules (~300–1000 m^2/m^3). The retentate gas can be introduced into another spiral would module to purify the gas to the desired levels. Gas flows to the next section of membrane modules to repeat this process until the necessary product gas specifications are met.

4.5.3. *Hollow fiber or tubular*

An alternative to the flat sheet membranes is the hollow fiber membranes. Compared to the flat-sheet membranes, hollow fiber membranes have several advantages including high membrane packing density, self-supporting structure and ease of handling. Hollow fiber membranes have several applications in gas separation and is one of the growing research area. Self-supporting hollow fibers are constructed in a microporous structure in order to have good mechanical properties and dense selective layer on the outer surface to separate the gases.

Typical hollow fiber membrane module with different parts and flow patterns is shown in Figure 4.9.[29] In the hollow fiber module, several thousands of fibers (combines several miles of fibers in length) must be packed into bundles and potted into casing.[30] An utmost care should be taken while preparing the hollow fibers and their corresponding modules to avoid the breakage of even a single fiber. Even a small defect in one hollow fiber can compromise the quality of the membrane module. Depending on the application, feed can be fed on the shell side or bore side while having permeate on the opposite side of the feed. For the shell side feed, hollow

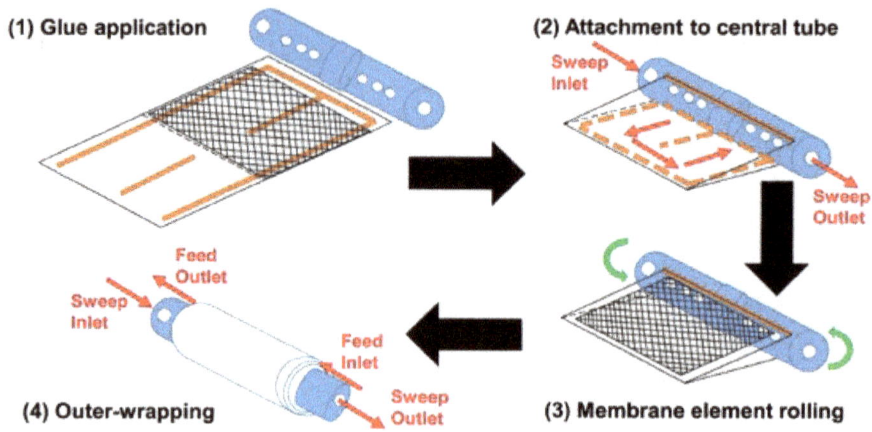

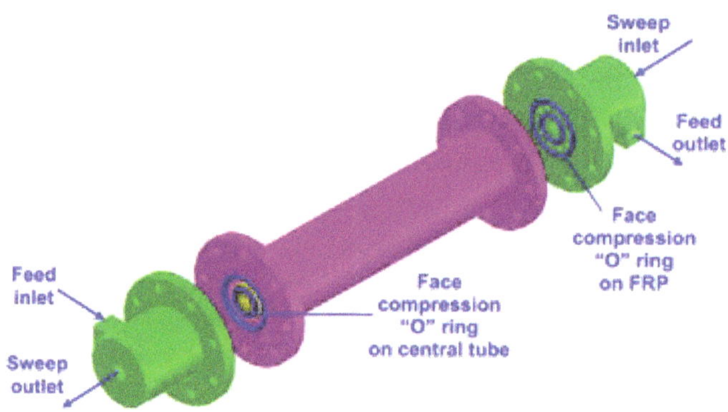

(a)

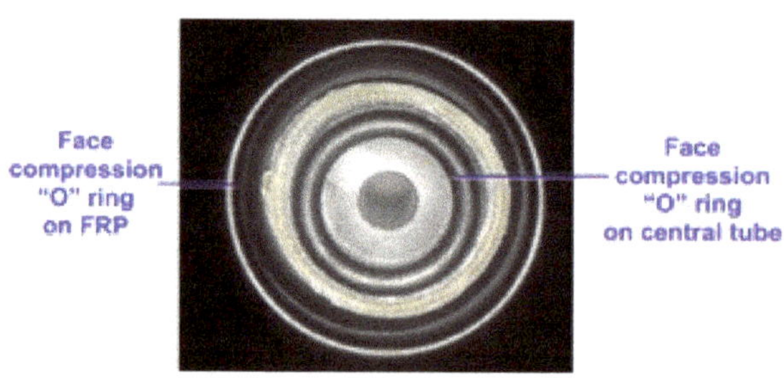

(b)

Figure 4.8. The step-by-step procedure of the spiral-wound membrane element fabrication. The face-compression "O" ring design of the spiral-wound membrane element with the Plexiglas FRP: (a) isometric view and (b) cross-section view. (Reprinted from Ref. 28 with permission from Elsevier.)

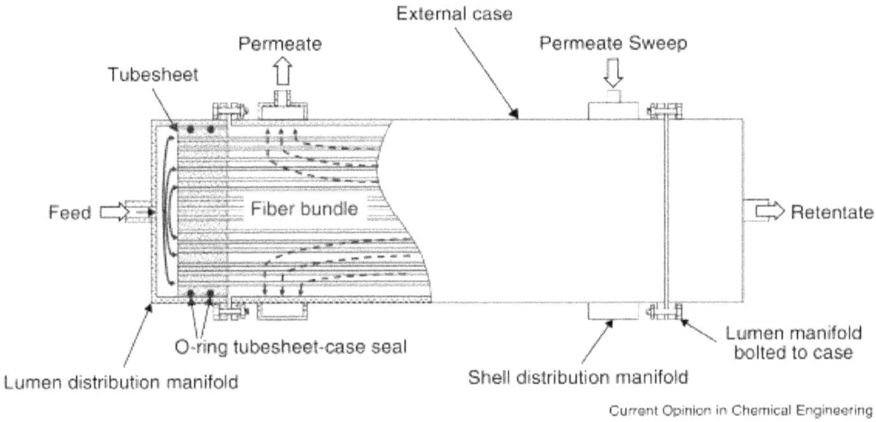

Figure 4.9. Cut-away view of a typical hollow fiber membrane module with a lumen feed and shell sweep operating in counter-current contacting mode. Key elements and flow paths are indicated: solid lines — lumen flow and dashed lines — shell flow. Note that the hidden portion of the module is a mirror image of that shown. (Reprinted from Ref. 29 with permission from Elsevier.)

fiber membranes can withstand to very high pressures and finds applications whose feed is available at higher pressures such as natural gas. Also the membrane area available for gas separation is also high compared to bore side feed. For the bore side feed, the feed pressure should be much lower than the shell side feed. There is always an intrinsic pressure drop in the small bore hollow fibers which can lead to the loss of driving force for gas separation. Hollow fiber modules can reach the highest packing density of 30,000 m^2/m^3 and is very high compared to spiral wound and plate-and-frame configuration as shown in Figure 4.10.

4.5.4. *Flow configurations*

The membrane modules are designed to operate mainly four different flow configurations; (1) dead end, (2) cross flow, (3) co-current, and (4) counter current as shown in Figure 4.11. Dead end configuration is mostly used in water purification and is a batch process. It is not a very effect method in the case of gas separation. In the case of cross flow configuration, feed gas and permeate flow direction have a 90° angle where feed gas is applied tangentially across the membrane surface.

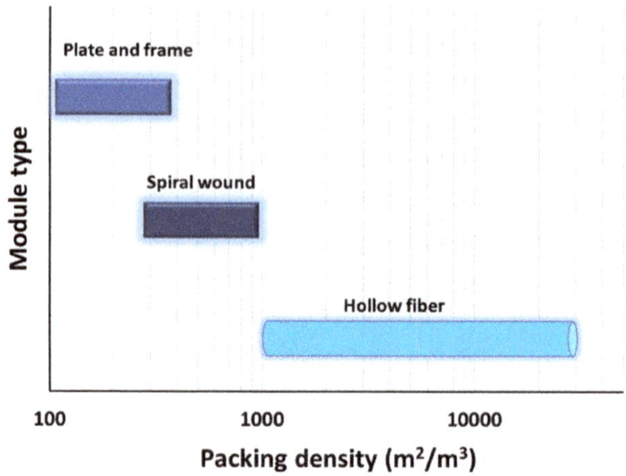

Figure 4.10. Comparison packing density of different membrane module configurations.

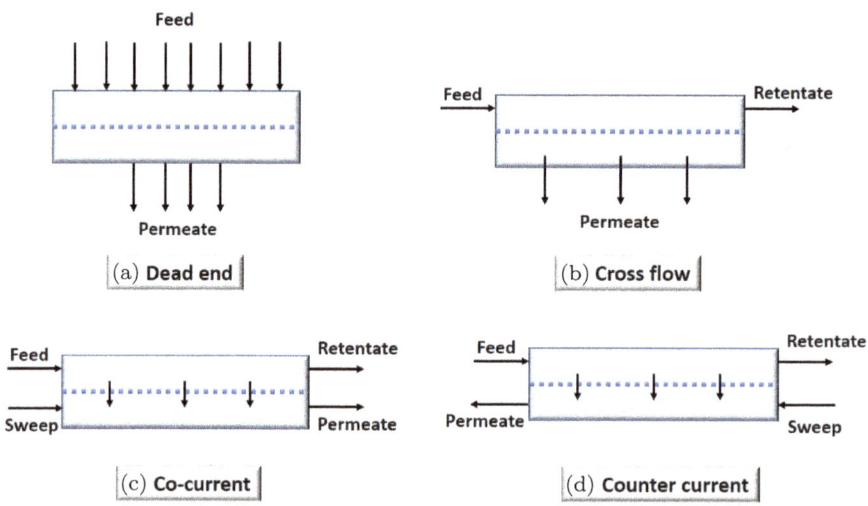

Figure 4.11. Different gas flow configurations in a membrane module for industrial processes.

Most gas separation applications involves either counter-current or co-current flow configurations. In the counter-current flow, the feed flows tangentially on the membrane surface and permeate gas sweep in the opposite direction of the feed as shown in the figure. In the co-current operation, the feed and the permeate flows in the same direction. In the co-current set up, the initial driving force in the co-current flow configuration is high. The driving force decreases as the gas permeates onto the other side and becomes equal after a certain length of the membrane resulting in the redundant membrane area. In contrast to co-current flow configuration, counter-current flow configuration maintains the driving force across the length of the fiber. Most industrial gas separation applications uses counter-current flow configuration because of its ability to maintain the driving force and having high gas separation efficiency of the membrane module.

References

1. Baker, R. W. (Richard W. & Wiley InterScience (Online service). *Membrane Technology and Applications*. (Wiley-Blackwell, 2012).
2. Koros, W. J. & Mahajan, R. Pushing the limits on possibilities for large scale gas separation: which strategies? *J. Memb. Sci.* **175**, 181–196 (2000).
3. Bong-Kuk Sea, *et al.* Hydrogen Recovery from a H_2-H_2O-HBr Mixture Utilizing Silica-Based Membranes at Elevated Temperatures. 1. Preparation of H_2O- and H_2-Selective Membranes. (1998). doi:10.1021/IE970730O
4. Doong, S. J. Membranes, adsorbent materials and solvent-based materials for syngas and hydrogen separation. in *Functional Materials for Sustainable Energy Applications* 179–216 (Elsevier, 2012). doi:10.1533/9780857096371.2.179
5. Venna, S. R. & Carreon, M. A. Amino-functionalized SAPO-34 membranes for CO_2/CH_4 and CO_2/N_2 Separation. *Langmuir* **27**, 2888–2894 (2011).
6. Venna, S. R. & Carreon, M. A. Synthesis of SAPO-34 crystals in the presence of crystal growth inhibitors. *J. Phys. Chem. B* **112**, 16261–16265 (2008).
7. Venna, S. R. & Carreon, M. A. Metal organic framework membranes for carbon dioxide separation. *Chem. Eng. Sci.* **124**, 3–19 (2015).
8. Denny Kamaruddin, H. & Koros, W. J. Some observations about the application of Fick's first law for membrane separation of multicomponent mixtures. *J. Memb. Sci.* **135**, 147–159 (1997).
9. Quan, S. *et al.* A bio-inspired CO_2-philic network membrane for enhanced sustainable gas separation. *J. Mater. Chem. A* **3**, 13758–13766 (2015).
10. Lin, H. & Freeman, B. D. Materials selection guidelines for membranes that remove CO_2 from gas mixtures. *J. Mol. Struct.* **739**, 57–74 (2005).
11. Venna, S. R. *et al.* Fabrication of MMMs with improved gas separation properties using externally-functionalized MOF particles. *J. Mater. Chem. A* **3**, 5014–5022 (2015).
12. Baker, R. W. Future directions of membrane gas separation technology. *Ind. Eng. Chem. Res.* **41**, 1393–1411 (2002).
13. Baker, R. W. & Low, B. T. Gas separation membrane materials: A perspective. *Macromolecules* 140930105103006 (2014). doi:10.1021/ma501488s

14. Galizia, M. *et al.* *50th Anniversary Perspective*: Polymers and mixed matrix membranes for gas and vapor separation: A review and prospective opportunities. *Macromolecules* **50**, 7809–7843 (2017).
15. Jansen, J. C. Ideal Separation Factor. in *Encyclopedia of Membranes* 1019–1020 (Springer Berlin Heidelberg, 2016). doi:10.1007/978-3-662-44324-8_302
16. Robeson, L. M. Correlation of separation factor versus permeability for polymeric membranes. *J. Memb. Sci.* **62**, 165–185 (1991).
17. Robeson, L. M. The upper bound revisited. *J. Memb. Sci.* **320**, 390–400 (2008).
18. Keizer, K., Uhlhorn, R. J. R., Van vuren, R. J. & Burggraaf, A. J. Gas separation mechanisms in microporous modified γ-al2o3 membranes. *J. Memb. Sci.* **39**, 285–300 (1988).
19. Holt, J. K. *et al.* Fast mass transport through sub-2-nanometer carbon nanotubes. *Science* **312**, 1034–1037 (2006).
20. Celebi, K. *et al.* Ultimate permeation across atomically thin porous graphene. *Science* **344**, 289–292 (2014).
21. Lively, R. P. Hollow fiber sorbents for post-combustion CO_2 capture. PhD Thesis, Georgia Institute of Technology, 2011.
22. Li, Y., Liang, F., Bux, H., Yang, W. & Caro, J. *J. Memb. Sci.* **354**(1–2), 48–54 (2010).
23. Baker, R. W. *Membrane Technology and Applications*. 2nd Edition. (John Wiley & Sons, Ltd, 2004). doi:10.1002/0470020393
24. Freeman, B. & Yampolskii, Y. *Membrane Gas Separation.* (Wiley, 2011).
25. Sircar, S., Rao, M. B. & Thaeron, C. M. A. Selective surface flow membrane for gas separation. *Sep. Sci. Technol.* **34**, 2081–2093 (1999).
26. Czichos, H., Saito, T. (Tetsuya) & Smith, L. R. (Leslie R. *Springer handbook of materials measurement methods* (Springer, 2006)).
27. Ryi, S.-K., Park, J.-S., Hwang, K.-R., Lee, C.-B. & Lee, S.-W. The property of hydrogen separation from CO_2 mixture using Pd-based membranes for carbon capture and storage (CCS). *Int. J. Hydrogen Energy* **38**, 7605–7611 (2013).
28. Salim, W. *et al.* Fabrication and field testing of spiral-wound membrane modules for CO_2 capture from flue gas. *J. Memb. Sci.* **556**, 126–137 (2018).
29. Mat, N. C., Lou, Y. & Lipscomb, G. G. Hollow fiber membrane modules, *Curr. Opin. Chem. Eng.* **4**, 18–24 (2014).
30. Hosseini, S. S., Dehkordi, J. A. & Kundu, P. K. Gas permeation and separation in asymmetric hollow fiber membrane permeators: Mathematical modeling, sensitivity analysis and optimization. *Korean J. Chem. Eng.* **33**, 3085–3101 (2016).

Chapter 5

Relevance of Molecular Gas Separations Through Membranes

5.1. Conventional techniques for gas separations

In general, most industrial relevant gas separations are challenging because of the similar size of the molecules to separate, and limited adsorption selectivity of the desired molecule to be separated, among other reasons. A typical example is nitrogen rejection from natural gas. N_2 rejection is a challenging technical separation because of the similar molecular size of N_2 (0.364 nm) vs CH_4 (0.38 nm). Furthermore, this particular separation lacks adsorption selectivity since there is CO_2 or H_2S, which greatly compete with N_2 and result in N_2 poor adsorption selectivity. Benchmark technologies to separate gas mixtures include distillation, pressure swing adsorption (PSA), lean oil absorption, and membrane separation.[1]

Distillation

Several gas mixtures are separated industrially through distillation. This process separate gas mixtures based on the boiling points difference between gas molecules. For example, most of the N_2 rejection plants in the US apply a cryogenic distillation process.[1] This process employs low temperature thermodynamic distillation based on the boiling points difference between CH_4 ($-162°C$) and N_2($-195°C$). As the mixture cools, CH_4 condenses before N_2, allowing the two gases to be separated. Cryogenic plants are suited to large gas fields that can deliver 50–500 million standard cubic feet per day (scfd) of gas for 10–20 years.[1] The main disadvantages of distillation processes are (i) high-energy requirement, (ii) high operation cost, and (iii) high tendency for blockage of process equipment.

Pressure Swing Adsorption (PSA)

In addition to distillation process, PSA has been employed to separate gases.[2] PSA is known to be one of the most economic and widespread processes for gas purification. Typically, PSA employs a zeolite absorbent or activated carbon to selectively adsorb specific gases (typically lower and higher hydrocarbons). The adsorbed gases are then regenerated from the absorbent with a combination of pressure and thermal changes to give a gas free of the molecule to be separated. However, there are drawbacks of this process, such as (i) limited gas stream adsorption, (ii) discontinuous circulation, (iii) complex design of the process and (iv) difficulty in scale up.

Lean Oil Absorption

Lean oil absorption technology absorbs a molecule away from another molecule and it regenerates the absorbed molecule either through staged pressure reduction or through thermally driven distillation.[3] This technique uses the difference in liquid-vapor equilibrium constants between the gas molecules to selectively absorb the gaseous molecules in the condensate at room temperature (RT) and high pressure, leaving the desired molecule to be separately unabsorbed. The liquid fraction containing the undesirable gas is then treated to give away the desirable free gas as a product. However, this technique is a less competitive alternative with respect to cryogenic process because of the low efficiency.[4]

Membrane Technology

Membrane separation offers the advantage of low-energy cost relative to the established gas separation processes. Membrane technology has been widely used in various industrial separation applications for the last two decades.[5] Currently, industrial relevant gas separations are dominated by polymeric membranes. Recently, there has been great interest in the development and application of inorganic, organic (carbon molecular sieve), and hybrid membranes because of their appealing properties for molecular gas separations. In the next section, we discuss the importance of separating gas mixtures through membrane technology.

5.2. Importance of membrane technology for molecular gas separations

Figure 5.1 illustrates the total energy consumption in the US by sector. Chemical separations account for about 50% of the US industrial energy use, and up to15% of the US total energy consumption.[6] Most of these industrially employed separations including distillation, evaporation, and drying are thermally driven. The energy consumed by chemical separation processes is about 45–55% in the industrial sector. Of this percentage, only about 20% correspond to a non-thermal separation. It has been reported that separation technologies that don't use heat could make industrial separations about 10 times more energy efficient.[6] In this respect, non-thermally driven membrane technology could play a key role in making industrially relevant gas separations less energy intensive and therefore economically feasible.

Energy efficient separation approaches in the petroleum, chemical and paper manufacturing sectors to purify chemicals could potentially save

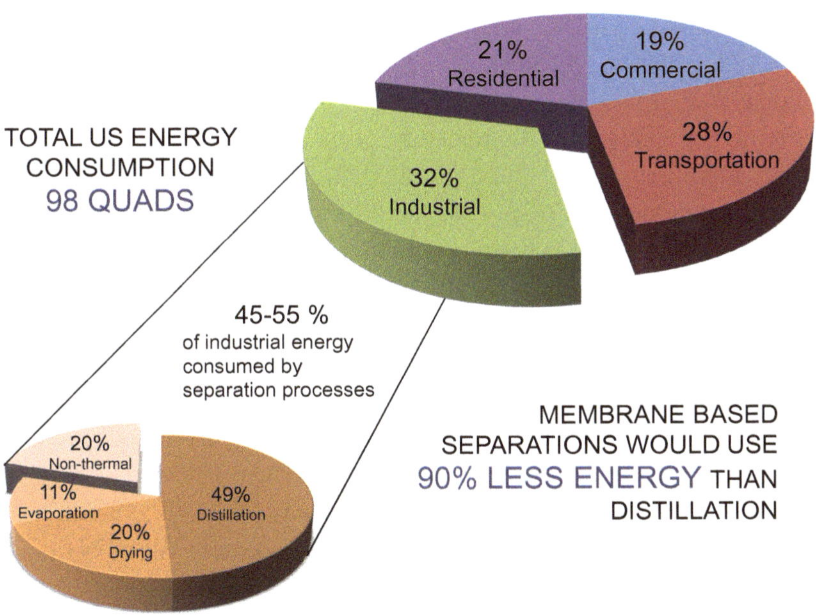

Figure 5.1. Total US energy consumption by sector, highlighting the importance of the energy consumed by chemical separations (modified from Ref. 6).

about 100 million tonnes of carbon dioxide emissions and US$4 billion in energy costs annually.[7]

5.3. Pre-economic evaluation: Membrane technology vs distillation

Very few reports exist in the open literature regarding economic evaluation of membrane technology vs benchmark distillation technology. An illustrative example of a preliminary economic evaluation of a membrane process for N_2 rejection was reported by Carreon, Yu and Li groups.[8] In this study, Aspen HYSYS software was used to model the membrane process for N_2/CH_4 separation. The major economic evaluation bases were: (i) $4 per MSCF gas price to plant, (ii) 4% N_2 in the residue gas, (iii) membrane module costs of $400/m^2 in a skid, and (iv) membrane lifetime of 15 years.[8] The purpose of this preliminary economic evaluation was to find out the required N_2 permeance and N_2/CH_4 selectivity for a *one-stage membrane process* to compete vs cryogenic distillation where the costs are ~$2.40/MSCF for a plant capacity of 1 MMSCF/day and $0.80/MSCF for a plant capacity of 20 MMSCF/day natural gas plant. These costs are typical costs for a 15 vol.% N_2 in the feed.

Figure 5.2 shows the N_2 rejection costs as a function of N_2 permeance for membranes with N_2/CH_4 selectivities of 8 and 15. The costs here include membrane processing cost and the revenue loss due to the methane loss to the permeate. For both selectivities, the costs decrease sharply as the N_2 permeance increases from 100 to 500 GPU. As the N_2 permeance increases further, the increased costs decrease more slowly. This preliminary economic evaluation indicated that the cost for SAPO-34 membrane (which display N_2 permeance of 500 GPU and selectivity of 8) would be approximately $1.13/MSCF if these membranes are used for N_2 rejection in the natural gas processing. This cost is less than half of the cryogenic distillation for a plant capacity of 1 MMSCF/day (~$2.40/MSCF).[8]

Figure 5.3 illustrates more cost sensitivity studies. When N_2 permeance is 500 GPU, an increase in N_2/CH_4 selectivity from our current 8 to 30 causes a decrease in cost by 45% (from $1.13/MSCF to $0.62/MSCF) for a N_2 feed concentration of 15 vol.%. Again, further increasing N_2 permeance warrants further reduction in N_2 rejection cost. The cost could

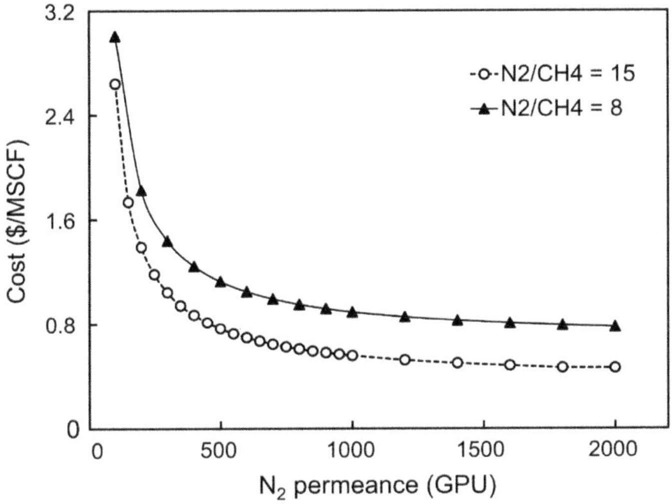

Figure 5.2. N_2 rejection cost as a function of N_2 permeance for membranes with N_2/CH_4 selectivities of 8 and 15. N_2 feed concentration is 15 vol.%. (Reprinted from Ref. 8 with permission from Elsevier.)

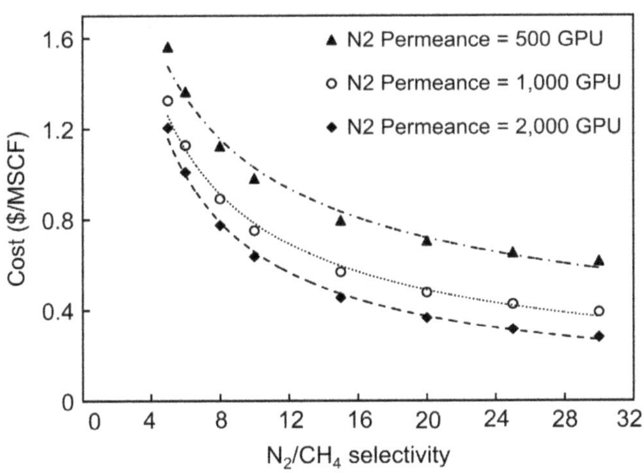

Figure 5.3. N_2 rejection cost as a function of N_2/CH_4 selectivity for membranes with N_2 permeances of 500, 1000, and 2000 GPU. N_2 feed concentration is 15 vol.%. (Reprinted from Ref. 8 with permission from Elsevier.)

be as low as \$0.28/MSCF for membranes displaying separation selectivity of 30 and N_2 permeance of 2,000 GPU.[8] This study indicates that for small nitrogen-contaminated gas wells, membrane technology (specifically SAPO-34 membranes) possess cost advantage over the benchmark technology cryogenic distillation.[8]

In a recent study, the economic evaluation indicates that the projected N_2 separation cost would be approximately \$0.60/MSCF for SAPO-34 membranes, if these membranes were to be used for N_2 rejection in the natural gas processing.[9] This cost represents only ~25% the benchmark technology cryogenic distillation for a plant capacity of 1 MMSCF/day (~\$2.40/MSCF) and 75% for a plant capacity of 20 MMSCF/day ((~\$0.8/MSCF).[9] It is important to mention that the economic cost comparison has been done for membranes operating at low feed pressures. Nevertheless, these studies clearly demonstrate the potential of membrane technology to compete and leapfrog distillation, the current benchmark technology to purify gases.

5.4. Representative reviews on MOF membranes and MOF mixed matrix membranes

In the next sections, we discuss research progress on MOF membranes and MOF mixed matrix membranes for several industrial relevant gas separations. These gas separations mainly include CO_2 capture, natural gas purification, noble gas separation, olefin/paraffin separations, and hydrogen purification. Comprehensive and representative reviews on the progress of MOF membranes for molecular gas separations have been reported elsewhere.[10–15] Reviews on MOF mixed matrix membranes are well documented too.[16,17]

Lin reviewed the stability of metal-organic framework membranes, as well as representative examples of separation properties of these membranes for gas and liquid separations.[10] This review highlights MOF membranes as a viable separation technology requiring fundamental understanding of their thermal and chemical stability, and the development of cost-effective synthesis of these membranes on suitable scalable supports.

Jeong's group reviewed progress on MOF membranes with an emphasis on their fabrication techniques, challenges involved in membrane synthesis, reported strategies to address these challenges, and gas separation performance.[11] Finally, this review offers perspectives on future research directions in this area.

Yao and Wang discuss research progress that has been made in developing zeolitic imidazolate framework (ZIF) membranes and thin films for gas separation, pervaporation and functional devices.[12] This review discusses developments of ZIF membranes/films, ZIF–polymer mixed matrix membranes and their functional applications.

Qiu *et al.* reported an overview of the diverse MOF membranes that have been prepared, focusing on the current techniques for the fabrication of MOF membranes.[13] Gas and liquid separation applications with different MOF membranes are also included in this review.

Venna and Carreon reviewed research progress on MOF membranes which have demonstrated ability to separate carbon dioxide from different light gases.[14] More specifically, this review focuses on gas separations which are relevant in flue gas treatment, natural gas purification, and hydrogen purification. The review discusses several conventional and novel strategies developed by several research groups for the continuous defect-free MOF membrane fabrication. Finally, the advantages of using MOFs in mixed matrix membranes and improvements in gas separation performances with the MOF-based mixed matrix membranes are presented.

Kang *et al.*'s review[15] focuses on recent advances in improving the performance of MOF membranes, involving the issues faced with MOF designation and growth for practical applications. Strategies for permeability and selectivity enhancement of MOF membranes, in terms of obtaining ultra-thin two-dimensional (2D) MOF nanosheets, and fine-tuning the pore size of the MOF framework are presented. The review also discusses the recent potential resolutions to the problems of MOF membranes for future practical applications including scale-up preparation and stability improvement.

Seoane *et al.* review the field of metal-organic framework-based mixed matrix membranes, with particular emphasis on their application in CO_2 capture.[16] Advances in the manufacture of these composite membranes on hollow fibers are discussed. Furthermore, this review illustrates efforts in understanding the separation performance of these composite materials.

Keskin's group discusses developments in MOF-based mixed matrix membranes by reviewing both the experimental studies and computational modeling methods that can accelerate selection of MOF/polymer combinations for high-performance gas separations.[17]

References

1. Baker, R. W. & Lokhandwala, K. Natural gas processing with membranes: An overview. *Industrial & Engineering Chemistry Research* **47** (7), 2109–2121 (2008).
2. Rufford, T. E., Smart, S., Watson, G. C. Y., Graham, B. F., Boxall, J., da Costa, J. C. D. & May, E. F. The removal of $CO_2$and $N_2$from natural gas: A review of conventional and emerging process technologies. *Journal of Petroleum Science and Engineering* 94–95, 123–154 (2012).
3. Sep-Pro Systems: Nitrogen Rejection Units. (2015). http://www.sepprosystems.com/Nitrogen_Rejection_Units.html
4. Carugati, A., Gambarotta, E. & Pollesel–EniTecnologie, P. (2005). Nitrogen rejection by absorption with field condensate. GTI — Natural Gas Technologies 2005.
5. Strathmann, H. Membrane separation processes: Current relevance and future opportunities. *AIChE Journal* **47** (5), 1077–1087 (2001).
6. Sholl, D. S. & Lively, R. P. Seven chemical separations to change the world. *Nature* **532**, 435–437 (2016).
7. US Dept. Energy Advanced Manufacturing Office. Bandwidth Study on Energy Use and Potential Energy Saving Opportunities in U.S. Petroleum Refining (US Dept. Energy, 2015).
8. Li, S., Zong, Z., Zhou, S. J., Huang, Y., Song, Z., Feng, X., Zhou, R., Meyer, H. S., Yu, M., Carreon, M. A. SAPO-34 Membranes for N_2/CH_4 separation: Preparation, characterization, separation performance and economic evaluation. *Journal of Membrane Science* **487**, 141–151 (2015).
9. Zong, Z. & Carreon, M. A. Thin SAPO-34 membranes synthesized in stainless steel autoclaves for N_2/CH_4 separation. *Journal of Membrane Science* **524**, 117–123 (2017).
10. Lin, Y. S. Metal organic framework membranes for separation applications. *Current Opinion in Chemical Engineering* **8**, 21–28 (2015).
11. Shah, M., McCarthy, M. C., Sachdeva, S., Lee, A. K. & Jeong, H.-K. Current status of metal-organic framework membranes for gas separations: promises and challenges. *Ind. Eng. Chem. Res.* **51**, 2179–2199 (2011).
12. Yao, J. & Wang, H. Zeolitic imidazolate framework composite membranes and thin films: synthesis and applications. *Chem. Soc. Rev.* **43**, 4470–4493 (2014).
13. Qiu, S., Xue, M. & Zhu, G. Metal-organic framework membranes: From synthesis to separation application. *Chem. Soc. Rev.* **43**, 6116–6140 (2014).
14. Venna, S. R. & Carreon, M. A. Metal organic framework membranes for carbon dioxide separation. *Chem. Eng. Sci.* **124**, 3–19 (2015).
15. Kang Zixi, Fan Lili & Sun Daofeng. Recent advances and challenges of metal-organic framework membranes for gas separation. *J. Mater. Chem. A* **5**, 10073–10091 (2017).

16. Beatriz Seoane, Joaquin Coronas, Ignacio Gascon, Miren Etxeberria Benavides, Oğuz Karvan, Jürgen Caro, Freek Kapteijn & Jorge Gascon. Metal-organic framework based mixed matrix membranes: A solution for highly efficient CO_2 capture? *Chem. Soc. Rev.* **44**, 2421–2454 (2015).
17. lknur Erucar, Gamze Yilmaz & Seda Keskin. Recent advances in metal-organic framework-based mixed matrix membranes. *Chem. Asian J.* **8**, 1692–1704 (2013).

Chapter 6

Metal-Organic Framework Membranes for CO_2 Separation

6.1. Introduction

From the environmental and energy view point, the separation and recovery of carbon dioxide (CO_2) from power generation, industrial operation and natural gas are of great interest. CO_2 is the main component of greenhouse gases, and its accumulation in the environment is leading to severe global warming issues. It has been estimated that the total CO_2 emissions in the US grew to 6,022 million metric tons (MMT) in 2007, which represents more than 80% of the total greenhouse emissions.[1] CO_2 emissions worldwide in 2005 were estimated at 28,051 MMT and projected to be 42,325 MMT in 2030.[2] In order to attenuate CO_2 accumulation, it is critical to separate and recycle carbon dioxide before it is released into the air. From the energy viewpoint, CO_2 is an undesirable impurity in natural gas wells, with concentrations as high as 70%. About 17% of all domestic natural gas in the US is treated to remove CO_2 before it is passed to the pipeline.[3] Pipeline specifications for natural gas require a CO_2 concentration below 2–3%. In addition, CO_2 must be separated from CH_4 because it reduces the energy content of the gas; CO_2 is also acidic and corrosive in the presence of water. Moreover, removing CO_2 without large energy expenditures is also desirable, and thus membranes that preferentially permeate CO_2 at high selectivities can significantly impact utilization of these gas wells by reducing the costs of natural gas purification.[4] Another relevant technology related to CO_2 capture involves the reaction of a fuel with oxygen or air to produce H_2 and CO (syngas). In the pre-combustion process, the formed CO reacts with steam producing a mixture of CO_2 and H_2 which is desirable to be separated to obtain pure hydrogen which can be used directly as a fuel source.

The benchmark technology used for CO_2 removal is amine adsorption, but amine plants typically lead to several issues.[3] First, capital costs associated to amine plants are high. Second, operation of these plants may be complex, and finally maintenance in these plants is expensive and labor intensive. The capture of CO_2 from flue gas and natural gas wells and from pre-combustion involves treating enormous gas volumes. In this respect, it has been proposed that membrane technology could play a pivotal role in making the separation process economically feasible. Membrane separation processes have several distinctive advantages over conventional amine adsorption.[5] For instance, membrane technology is a viable energy-saving method, since it does not involve any phase transformation. Furthermore, the required membrane process equipment is simple, easy to operate, control and scale-up.

Metal-organic frameworks in membrane form, have emerged as an appealing type of crystalline microporous materials which combine highly desirable properties, such as uniform micropores, high surface areas, and exceptional thermal and chemical stability, making them ideal candidates for CO_2 gas separations. In this chapter, we review research progress on metal-organic framework membranes which have demonstrated ability to separate carbon dioxide from different light gases. We have focused mainly on CO_2/N_2, CO_2/CH_4, and CO_2/H_2 gas separations which are relevant binary compositions in flue gas treatment, natural gas purification, and hydrogen purification, respectively.

6.2. Justifying the use of MOF membranes for CO_2 separation

One may ask, why MOFs as membranes for CO_2 separation? Some of the most key reasons are described in the next paragraphs. Due to its remarkable high CO_2 uptakes, open porous framework structure with large accessible pore volumes, chemical stability in the presence of hydrocarbons and water (some of them), and limiting pore apertures in the range of the kinetic diameters of several relevant gas molecules, MOFs are highly appealing materials for CO_2 capture. CO_2 adsorption capacities of several potential MOFs compared to zeolites is shown in Table 6.1. It is clear that most MOFs display higher CO_2 uptakes as compared to traditional zeolite phases.[6–14]

In particular, zeolitic imidazolate frameworks (ZIFs), a subclass of MOFs, have emerged as a novel type of crystalline porous materials which

Table 6.1. CO_2 adsorption capacity of several MOFs compared to representative zeolites.

Porous crystal	Temperature (K), Pressure (bar)	CO_2 adsorption (mol/kg)	Reference
HKUST-1	298, 35	10.7	6
MIL-53	304, 25	10	7
MIL-100	304, 50	18	8
MOF-177	298, 35	33.5	6
MOF-5	298, 35	21.7	6
IRMOF-3	298, 35	18.7	6
MIL-101	304, 50	40	8
ZIF-8	293, 1	2.6	9
SAPO-34	293, 1	3	10
DDR	198, 1.85	2.8	11
H-ZSM-5	281, 0.81	2.15	12
Zeolite 13X	298, 21	5.2	13
Zeolite 5A	303, 10	3.55	14

combine highly desirable properties from both zeolites and MOFs, such as microporosity, high surface areas, and exceptional thermal, chemical stability, making them ideal candidates for gas separation applications.[15] In ZIFs, metal atoms such as Zn, Co, and Cu are linked through N atoms by ditopic imidazolate (Im) or functionalized Im links to form neutral frameworks and to provide tunable nanosized pores formed by 4 and 6-ring ZnN_4, CoN_4, and CuN_4 clusters. The framework of ZIF compounds closely resemble the framework of zeolites, i.e the T-O-T bridges (T = Si, Al, P) in zeolites are replaced by M-Im-M bridges (M = Zn, Co, Cu) and coincidentally, their bond angles in both structures are 145°.[16] Therefore it is not surprising that most of the documented MOF membranes for CO_2 separation have ZIF compositions.

Metal-adeninate biometal organic frameworks, known as Bio-MOFs[17–22] have been identified as appealing candidates for CO_2 separation because of its permanent microporosity, high surface areas, chemical stability, and due to the presence of basic bio-molecule building units which promote preferential CO_2 adsorption, resulting in exceptionally high CO_2 adsorption capacities. A prototypical type of metal-adeninate biometal organc framework is Bio-MOF-1 ($Zn_8(ad)_4(BPDC)_6O \cdot 2Me_2NH_2$; where ad = adeninate, BPDC = biphenyldicarboxylate), a three-dimensional porous metal-organic framework with infinite zinc-adeninate columnar secondary building units (SBUs) which are interconnected via biphenyldicarboxylate linkers.[16] Interestingly, the pore size of these BioMOFs can be adjusted by choosing suitable organic linkers.[22] A comprehensive review on CO_2 adsorption and separation over MOFs has been documented elsewhere.[23] In this review, it was concluded that although many MOFs have been synthesized and characterized structurally, only a small fraction of MOFs have been checked for CO_2 adsorption, storage, and gas separation.[23] The field of MOF-based membranes for CO_2 capture has high potential for rapid growth because of the high number of available MOFs, their pore tunability that could lead to molecular sieving properties, and most importantly their high CO_2 uptakes.

6.3. Strategies for MOF membrane fabrication

In this section, we summarize representative synthesis strategies that have been successfully employed to prepare continuous MOF membranes. We have reviewed these approaches elsewhere.[24] The successful fabrication of continuous high quality MOF membranes is challenging due to the low heterogeneous nucleation density, difficulty to form direct chemical bonds between the organic linkers of MOFs and the support surface, relative weakness of the coordination bonds and intergrowth of MOF crystals, among other reasons.[25] Furthermore, membrane cracking may develop during the membrane activation step or solvent removal step due to the differences in thermal expansion coefficients between the MOF layer and the support. Several research groups have succeeded in preparing dense defect-free MOF membranes by facilitating the nucleation and growth of MOF crystals using several synthetic protocols including *in situ* growth,[26,27] surface modification with organic functional groups,[28,29] use of electrostatically compatible supports,[30] using microwave reactor[31,32] mainly. Shah *et al.* summarized the different MOF fabrication techniques in detail and provided a lot of useful information to improve the present membranes.[33] In this chapter, some of the novel methods developed recently are discussed in detail. Some of the above mentioned approaches were borrowed from zeolite membrane fabrication techniques. Other more sophisticated techniques include liquid phase deposition,[34] cathodic electrodeposition,[30] and contra-diffusion approach.[35]

6.3.1. *In situ growth*

In this approach, the membrane grows directly on a support which is not modified with any pre-made seed particles. In this method, nucleation and growth occurs at the same time. Using the *in situ* solvothermal method, first continuous and dense thin-film MOF membranes were fabricated on α-alumina supports by Liu *et al.*[36] The membranes were well inter-grown with randomly packed grains without cracks. Most importantly, the membranes had good adhesion with the α-alumina supports. In another study on ZIF-69 membrane fabrication by Liu *et al.*, the orientation of ZIF-69 membranes was effectively controlled along *c*-axis using *in situ* method on α-alumina supports.[37] Caro and coworkers prepared ZIF-90 membranes using 3-aminopropylsilane as a covalent linker between the ZIF-90 layer and the alumina support via imine condensation. First, ethoxy groups of silane react with the –OH groups of the support and during the solvothermal synthesis the amino groups reacted with aldehyde groups of organic linker followed by the nucleation and growth of the membranes.[38,39] In other work by Xie *et al.*, the alumina particles were first modified with 3-aminopropylsilane and then deposited on the coarse macroporous alumina support to reduce the pore size of the support and increasing nucleation and crystal growth density. The resultant $\sim$2 μm defect-free membrane displayed moderate gas separation selectivities.[29]

Zeolitic Imidazolate Framework (ZIF-8) membranes were also grown on the titania supports using microwave assisted solvothermal synthesis.[31] Bux *et al.* used a modified ZIF-8 synthesis protocol by replacing DMF with methanol as the solvent to prepare these membranes. The membranes were continuous and EDXS analysis clearly showed the attached between the support and ZIF-8 layer. Jeong and coworkers reported an interesting one step *in situ* synthesis method to fabricate the ZIF-8 membranes on unmodified porous α-alumina supports in the presence of sodium formate.[40] Sodium formate played a critical role by enhancing the heterogeneous nucleation and intergrowth of ZIF-8 crystals on supports. The continuous defect-free films of other ZIFs such as ZIF-7, ZIF-90, and SIM-1 have also been prepared using this method in the presence of sodium formate.[40] ZIF-8 membranes were also prepared on the polysulfone supports using the similar method and in the presence of sodium formate by Cacho-Bailo *et al.* and carried out an alternate crystallization method to produce micro and nanometer sized crystals on the support, which avoided holes and defects in the ~35 μm thick membrane.[41] Nagaraju *et al.* also prepared ZIF-8 and CuBTC membranes on the porous asymmetric polysufone membrane by *in situ* growth followed by layer by layer deposition method without using any sodium formate and without any seeding.[26]

In order to increase the flux of MOF membranes and decrease the volume of the modules, hollow fiber supports have been greatly preferred for MOF membrane synthesis. The MOF membranes were also fabricated on both inorganic and polymeric tubular or hollow fiber supports using the *in situ* growth method. A substituted imidazolate-based MOF (SIM-1) membrane was fabricated inside a long tubular asymmetric alumina support using the one step *in situ* growth method.[42] Aguado *et al.* used more polar and bulkier organic linker to synthesize the MOF and to fabricate the membranes which also resulted in the decrease of the pore size.[42] Xu *et al.* prepared ZIF-8 membranes on the outer surface of unconventional alumina hollow fiber support using a concentrated synthesis gel.[43] Highly viscous concentrated gel was easily deposited on the surface of the hollow fibers and resulted in the rapid ZIF-8 nucleation and growth, leading to intergrowth of the crystals. Lai and coworkers also used similar method to prepare the ZIF-8 membranes on hollow yttria-stabilized zirconia supports by using the seeding method.[44]

Several research groups have attempted to fabricate these membranes on cheaper polymer supports in order to minimize the overall cost of the MOF membranes. Uniform ZIF-8 membrane was grown on the inner surface of the APTES modified ceramic hollow fiber using a modified seeding process, termed as "cycling precursors method", in which the precursors continuously flow through the lumen of the support.[45] A modified *in situ* seeding process was used to prepare a uniform ZIF-8 seed layer and fabricated the ZIF-8 membrane on the inner-side of the hollow fiber support.[45] Continuous and defect-free HKUST-1 films on the surface of polyvinylidenefluoride (PVDF) hollow fibers were rapidly fabricated using pressure-assisted room temperature growth strategy by using solid copper hydroxide nanostrands as the copper source.[46] A copper hydroxide nanostrand layer was precoated on the PVDF hollow fiber and then continuous HKUST-1 membranes were grown by immersing the precoated PVDF hollow fiber into a H_3 BTC ethanol/water solution at room temperature. Continuous polycrystalline ZIF-90 membranes have been grown at 65°C on nanocrystal-seeded surfaces of poly(amide–imide) macroporous hollow fiber supports by liquid phase process.[47] Nair and coworkers introduced a novel method, called interfacial microfluidic membrane processing, to fabricate continuous molecular sieving ZIF-8 membranes as poly(amide-imide) hollow fiber supports. Film growth was directed on the fiber inner or outer surfaces through different reagent and solvent permutations.[48] This work effectively demonstrates that MOF membranes can be fabricated by a technologically scalable low temperature process on polymeric hollow fiber supports, with complete surface coverage to fabricate high-surface-area membrane modules.[48]

6.3.2. *Secondary growth*

Although the microstructures of the membranes produced by *in situ* growth method are well intergrown and continuous, most of these membranes have displayed only limited to moderate separation properties. Different from the *in situ* growth method, the secondary growth approach separates the nucleation and growth steps. In the *in situ* method, MOF crystals are easier to crystallize and grow in the bulk solution rather than on the support surface resulting in membranes having high concentration of defects. Secondary seeded growth approach has been developed to enhance interaction between the seed layer and the support such as microwave-assisted synthesis,[32] thermal seeding,[49] reactive seeding,[50] and pre-modification of the support surface by organic polymers.[51]

Liu *et al.* used the dip coating technique to seed the alumina support with the ZIF-69 particles.[52] Secondary seeded growth of the ZIF-69 membranes successfully resulted in continuous *c*-oriented membranes which was not possible with the *in situ* growth method under the same conditions. We previously reported continuous and reproducible ZIF-8 membranes fabricated using the secondary seeded growth method.[53] In our work, the inside support was rubbed with ZIF-8 seeds which provided the nucleation sites for membrane growth.[53] Solvothermal treatment resulted in continuous $\sim$5–9 μm membranes. NH_2-MIL-53(Al) membranes have been prepared on a macroporous glass frit discs by seeding the support with seed colloidal solution, followed by solvothermal route.[54]

Jeong and coworkers reported the fabrication of MOF-5 membranes using microwave induced thermal deposition (MITD) method.[32] This group first coated the alumina supports with graphite for better interaction with the microwave photons and then seeded with the MOF-5 crystals using microwave induced thermal deposition. During the secondary growth, N-ethyldiisopropylamine was used in precursor solutions which helped to prevent the seed detachment and to form defect-free continuous and preferentially oriented films. McCarthy *et al.* modified the support with the organic ligand by thermal seeding method and rapidly fabricated the ZIF-8 and ZIF-7 membranes by solvothermal method. The ligand solutions were dropped on the support and heated at 200°C for 20 min in order to form covalent bonds between the support and the organic ligand.[55]

Ranjan *et al.* prepared microporous metal-organic framework (MMOF) membranes on porous alumina supports using the secondary growth method as shown in Figure 6.1.[51] The surface of the alumina was first modified using polyethyleneimine (PEI) and then the MMOF seeds were deposited

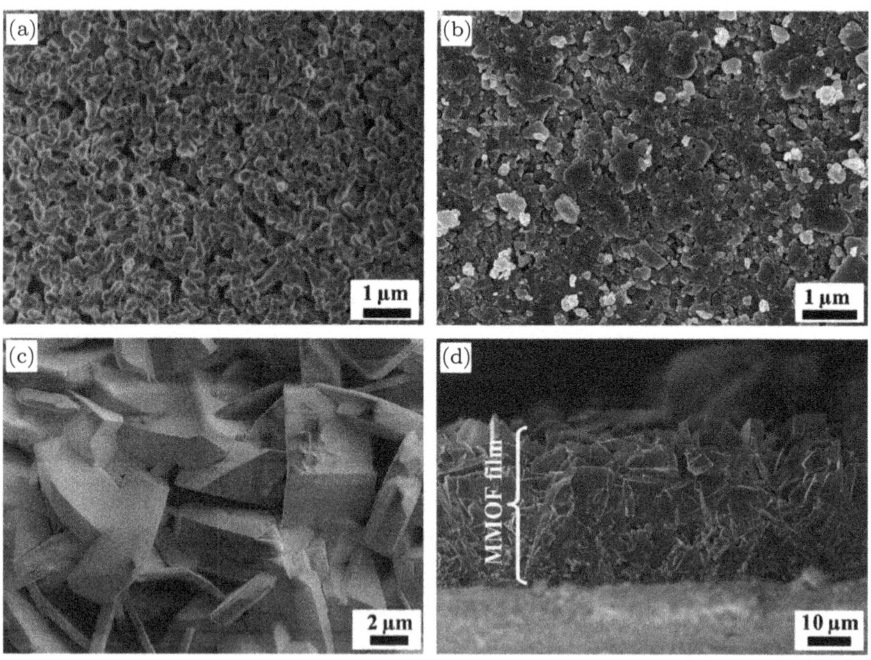

Figure 6.1. SEM images at different stages of MMOF membrane growth: (a) α-alumina support, (b) seed layer, (c) membrane (top view), and (d) membrane (cross-section view). (Reprinted from Ref. 51 with permission from American Chemical Society.)

manually on the modified supports. Ranjan *et al.* also controlled the orientation of membranes in *b*-direction and attributed to faster growth along the *b*-direction during the secondary growth with some misoriented crystals. Caro and coworkers[56] used similar seeding technique using polyethyleneimine (PEI) as the polymer binder, but the supports were seeded in a single step by dipcoating in an aqueous seeding sol which contained 4 wt.% ZIF-7 nano seeds and 2.5 wt.% polyethyleneimine (PEI). Microwave assisted secondary growth was carried out to crystallize a compact randomly oriented ZIF-7 layer and obtained a thin membrane.[56,57] Zhou *et al.* used chitosan as a binder instead of polyethyleneimine (PEI) to fabricate $Cu_3(BTC)_2$ membranes.[58] The researchers used the dip coating technique to seed on the α-alumina hollow ceramic fiber surface assisted with chitosan. Chitosan was used to improve the mutual interaction between the support and MOF owing to its abundance of amino and hydroxyl groups.[58]

Hu *et al.* developed a novel reactive seeding (RS) technique to fabricate the defect-free membranes and also studied the seed layer formation mechanism.[59] In this method, the seed layer was grown on the support that acts as a metal source by the reaction between the inorganic support and the organic precursor. The MIL-53 membrane was prepared using this reactive seeding method and achieved uniform, thin and well intergrown MIL-53 membranes.[59] Later, several research groups employed this reactive seeding method to prepare several continuous MOF membranes including ZIF-78,[59] ZIF-71,[60] Zn-BLD,[61] and MIL-96[50] on zinc oxide or alumina supports as the metal source.

High quality ZIF-8 membranes on alumina supports were prepared using aqueous solutions instead of organic solvents via seeded growth method.[62] This group reported that the basic pH of the aqueous synthesis solutions resulted in the higher depronated 2-methylimidazole as compared to the methanol-based synthesis solution. This led to well intergrown ZIF-8 membranes, since the organic ligands on the growing surface deprotonated before they could grow further and growth occurring in more directions to achieve better intergrowth.[62]

6.3.3. *Alternative MOF membrane synthesis approaches*

Bétard *et al.* reported that MOF membranes on macroporous alumina and titania supports using *stepwise liquid phase deposition of reactants*.[34] This method has several advantages for thin film preparation as compared to the solvothermal route. The main advantages are better control of the layer formation and reduced stress in the support/film boundary region. After 90 cycles of deposition, MOF crystallites grew $\sim 30\,\mu$m deep inside the support pores with some deposition on the support surface.[34] MOF membranes were also fabricated on polymer supports using layer by layer deposition. For example, Nagaraju *et al.* prepared the CuBTC and ZIF-8 membranes using *in situ* method followed by layer by layer deposition on the polysulfone supports and resulted in a continuous MOF membranes with the thickness of 10–25 μm.[26] Guo *et al.* prepared the Cu_3BTC_2 membranes supported on a copper mesh by oxidizing the mesh to form nucleation sites. The copper on the mesh and in the solution provided the metal source for crystal growth (twin copper source method) and formed a defect-free membrane.[63] A similar method was used to fabricate ZIF-8 membranes inside alumina tubular supports by modifying the support with ultrathin ZnO layer which acted as nucleation sites to form crystals and promote membrane growth.[64]

Oliver and coworkers prepared thin film of MOFs using zinc oxide nanowire modified substrates.[65] This method involves growing the zinc oxide nanowires on a transparent conductive oxide substrate and then exposed to the precursor solution, resulting in complete coverage of the nanowire substrate with interconnected crystals.[65] Li *et al.* fabricated highly oriented ZIF-7 nanorod membranes by tailoring the crystal size and morphology of the ZIF-7 using a random oriented crystal seeding.[66] Recently, Li *et al.* introduced a new approach called "cathodic electrodeposition method" to fabricate thin films of MOFs. Depending on the applied deposition potential at different concentrations of triethylammonium, two different MOF phases were formed and selectively prepared submicron thick mixed layer and bilayer films.[30]

Yao *et al.* developed a novel approach, named contra-diffusion method, to prepare the MOF membranes on a flexible nylon substrate.[35] Different synthesis solutions, metal salt solution and organic ligand solution were separated by a porous nylon membrane and the crystallization took place on the membrane surfaces through solution contra-diffusion. In principle, this method offers an easy and effective way to directly grow MOFs on flexible polymer substrates.[35] CuBTC membranes within porous supports were also fabricated using a counter-diffusion method, a method similar to

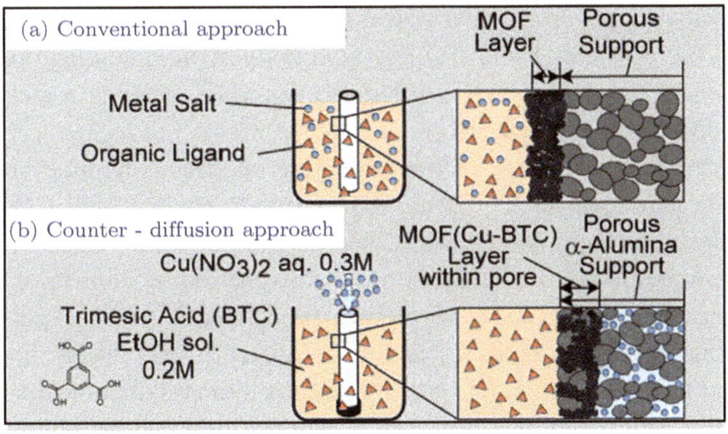

Figure 6.2. Schematic illustration of (a) the commonly used method to prepare MOF-based separation membranes and (b) the counter-diffusion procedure used in this study. (Modified from Ref. 67.)

contra-diffusion method. The method involves the supply of the reactants from opposite sides of the support and reactions occurs only within the support pores and continue until all the pores are plugged without any defects as shown in Figure 6.2.[67]

6.4. MOF membranes for CO_2 separation from flue gas

In the next paragraphs, we discuss the state of the art of the diverse MOF membranes that have been used for: CO_2/CH_4, CO_2/N_2, and CO_2/H_2 gas separations. These binary gas mixtures are relevant in natural gas purification, flue gas treatment, and hydrogen purification respectively. We compare the separation performance of MOF membranes vs zeolite and polymeric membranes.[24]

It is important to develop new advanced CO_2 capture technologies in order to maintain the cost-effectiveness of US coal-fired power generation. Compared to amine absorption, membrane processes require less energy to operate, and do not need chemicals or regenerating absorbents.[5] Also, membranes are compact and can be retrofitted onto the tail end of power-plant flue gas streams without complicated integration schemes. Published system analysis and feasibility studies demonstrate that membranes are a technically and economically viable option for CO_2 capture from flue gas exhaust in pulverized coal power plants.[68] For CO_2 capture from flue gases, CO_2 permeance and CO_2/N_2 selectivity at the intended operating environment are the two basic criteria to determine whether a membrane can be effectively utilized.

Polymeric membranes such as cellulose acetate, polyimide and polyaramide can separate CO_2 from N_2 mixtures, but the high CO_2 pressures plasticize them and decrease their separation ability considerably.[69] Polymeric membranes containing poly (ethylene oxide) units have shown moderate to good CO_2/N_2 separation properties.[70,71] Other polymeric membrane compositions have been summarized in the revised version of Robeson plot.[72] Typically, these polymeric membranes are not suitable for high temperature and high humidity conditions required for CO_2 capture from flue power plants.[73] Recently, amine carriers functionalized cross-linked poly(vinyl alcohol) membranes have shown remarkable separation performance for CO_2/N_2.[74] Although the current membrane technology is dominated by polymers, these membranes may suffer from low CO_2 permeability and selectivity, because they are based on a solution-diffusion mechanism, relying its separation ability mainly on the small size differences of the gas molecules.[73,75]

Zeolite membranes offer significant and distinctive advantages over polymeric membranes. For instance, they do not deform under high CO_2 pressures, have high thermal, mechanical, and chemical stability, and good erosion resistance.[76] The molecular sieving ability of zeolite membranes is well known and has been exploited for a vast number of relevant gas mixture separations,[77–81] including CO_2 separation from N_2.[82–87] These zeolite membrane compositions include zeolite-Y,[87] zeolite-T,[83] ZSM-5,[85] SAPO-34,[84] DD3R[86] and AlPO-18.[82] Because CO_2 (~ 0.33 nm kinetic diameter) is smaller than N_2 (~ 0.36 nm), some of these small-pore zeolites, whose pore sizes are close to that of CO_2 molecules, can separate CO_2/N_2 mixtures. Typically these zeolite membranes have moderate to high CO_2/N_2 selectivities due to a combination of differences in diffusivity and competitive adsorption, but only moderate CO_2 permeability.

Metal-organic frameworks,[15,88,89] have emerged as a novel type of crystalline porous materials which combine highly desirable properties, such as uniform micropores, high surface areas, and exceptional thermal and chemical stability, making them ideal candidates for gas separation applications.

Although the separation ability of several MOF compositions have been demonstrated for several gas pairs[31,37,53,56,57,62,66,90–93] including CO_2/CH_4, H_2/CH_4, H_2/N_2, H_2/O_2, H_2/CO_2,CO_2/CO, only few reports exist on the separation ability of MOF membranes for CO_2/N_2.[37,51,52,94] In general, low to moderate CO_2/N_2 selectivities have been observed for these MOF compositions.

Figure 6.3 compares the performance of (a) zeolite membranes and (b) MOF membranes for CO_2/N_2 separation. Zeolite membranes display much higher CO_2/N_2 separation selectivities as compared to MOF membranes. This may be related to the sharp molecular sieving effect of zeolites (due to its "rigid crystallographic pore size") as compared to the flexible MOF structure due to the presence of organic linkers leading to the well-known gate opening or breathing effect which limits the molecular sieving phenomena in MOFs. On the other hand, CO_2 fluxes (permeances) are in general higher for MOF membranes which is expected due to (at least in part) the flexible framework nature, and higher surface areas of MOFs. For clarity the zeolite and MOF compositions showing the highest CO_2/N_2 selectivity and CO_2 permeance are shown in Figure 6.3. For example in Figure 6.3a, zeolite Y shows the highest CO_2/N_2 separation selectivity with low CO_2 permeance.[87] Due to the 0.73 nm pore size of zeolite Y, there is no steric hindrance for CO_2 or N_2 to enter the pores. However, for CO_2,

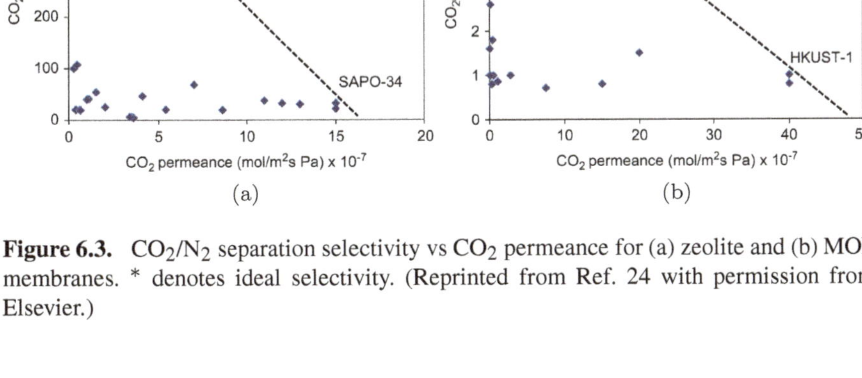

Figure 6.3. CO_2/N_2 separation selectivity vs CO_2 permeance for (a) zeolite and (b) MOF membranes. * denotes ideal selectivity. (Reprinted from Ref. 24 with permission from Elsevier.)

there is preferential interaction with the aluminosilicate framework and the extra-framework cations which leads to a favorable combination of sorption and mobility. Therefore, CO_2 permeates preferentially through the zeolite by surface diffusion, while N_2 is excluded from the zeolite Y pores due to preferential CO_2 adsorption and pore constriction, leading to the high observed CO_2/N_2 separation selectivities. In addition, the nanometer sized zeolite Y reported by the authors[87] may have contributed to the formation of close-packed arrays of crystals which led to a well inter-grown membrane. SAPO-34 [84] shows the highest CO_2 permeance, and moderate separation selectivity as shown in Figure 6.3a. SAPO-34 is a silicoalumino phosphate with 0.38 nm pore size which in principle allows for CO_2 molecular sieving over N_2. Authors explained the high CO_2 permeance to the formation of a high quality (low concentration of nonselective pores) thin membrane.

In the case of MOF compositions, as shown in Figure 6.3b, ZIF-69, a zeolitic imidazolate framework with pore size of 0.78 nm shows the highest CO_2/N_2 separation selectivity.[37] The moderate selectivity was attributed to the surface diffusion due to the adsorption affinity of CO_2 to ZIF-69. HKUST-1 a Cu-based MOF ($Cu_3(BTC)_2(H_2O)_3 \cdot xH_2O$ (where BTC = benzene 1,3,5-tricarboxylate)), with an interconnected 3D pore structure with pore size of 0.9 nm shows the highest CO_2 permeance,[95] which was attributed to the large pore size and large pore volume of the framework. Figure 6.4 shows representative SEM images of ZIF-69 membrane that have shown the highest ideal separation selectivity performance for CO_2/N_2 gas mixtures.

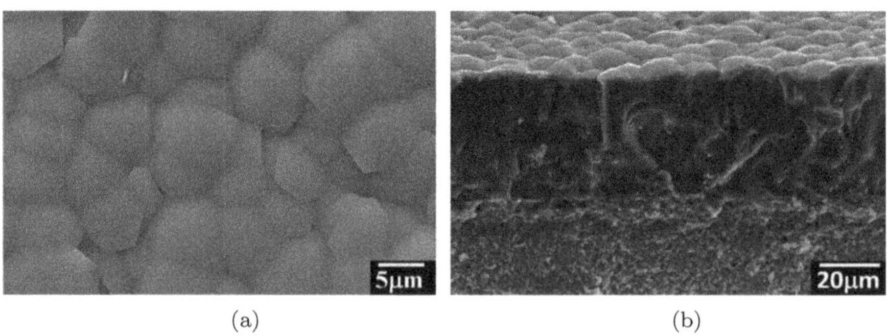

Figure 6.4. ZIF-69 membrane (a) top view and (b) cross section view employed to separate CO_2/N_2 gas mixtures. (Reprinted from Ref. 37 with permission from Elsevier.)

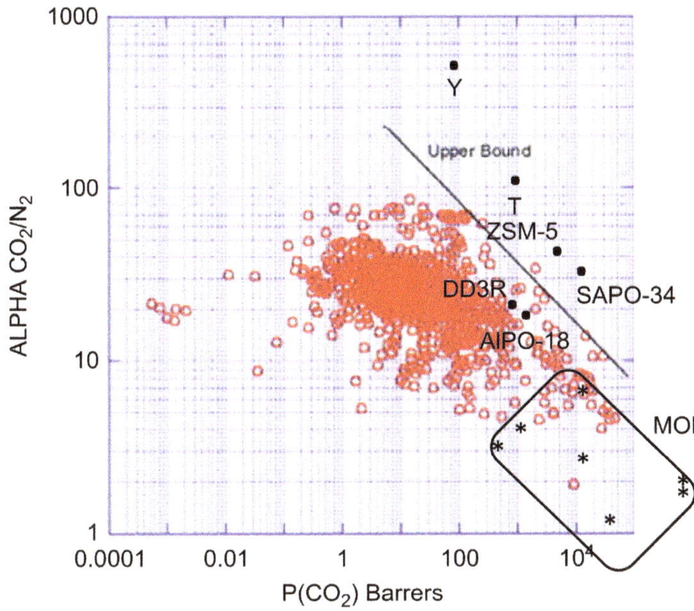

Figure 6.5. Robeson plot for CO_2/N_2 gas mixtures. Red points represent experimental points for polymeric membranes. For comparison zeolite membranes are included. CO_2 permeabilites for the two points shown out of the graph are $>10^6$ Barrers. (Reprinted from Ref. 24 with permission from Elsevier.)

The Robeson plot for CO_2/N_2 separation selectivity versus CO_2 permeabilities (permeance $\times$ membrane thickness) of polymeric membranes[72] has been widely used to compare the performance of membranes. Figure 6.5 compares the separation performance of polymeric, and zeolite membranes for CO_2/N_2 gas mixtures. Several zeolites, including Y, T, ZSM-5 and SAPO-34 perform much better (above the upper bound) than any polymeric membrane (red points). Although several zeolite membranes perform much better than any other membrane composition (including MOF compositions), still the CO_2 permeability needs to be improved for these membranes to be economically a viable option for CO_2 capture from flue gas. Merkel *et al.* showed the optimal membrane CO_2/N_2 selectivity for separation of CO_2 from flue gas can be as low as 30–40.[96] However, the CO_2 permeance must be much greater than $\sim 6.6 \times 10^{-7}$ mol/m$^2 \cdot$s$\cdot$ Pa ($\sim$2000 GPU). 1 GPU $= 10^{-6}$ cm^3(STP)/(cm$^2 \cdot$s$\cdot$cmHg). The commercially available CO_2-selective polymer membranes, however, only have a permeance of 100 GPU[96] which is too low for flue gas CO_2 capture. For MOF membranes, the performance lies below the upper bound. The cost of the MOF membrane fabrication is high as compared to typical polymeric membranes. In order to compete with polymeric membranes, MOF membranes should display high CO_2 permeance and high CO_2/N_2 selectivity in order to compensate with the cost of the MOF membrane fabrication.

6.5. MOF membranes for CO_2 separation from natural gas

Processing of natural gas is by far the largest industrial gas separation application.[97] Every year, close to 100 trillion standard cubic feet of natural gas are used worldwide. The US consumption of natural gas accounts for ~22 trillion standard cubic feet per year. As mentioned previously, natural gas needs treatment before it enters the pipeline (removal of CO_2). Polymeric membranes have been used to separate CO_2 from natural gas, and these membranes include cellulose acetate,[98] polyimide,[99] and perfluoro-based[100] polymer membranes. Similar to flue gas treatment, polymeric membranes can separate CO_2/CH_4 mixtures, but the high CO_2 pressures can plasticize them and decrease their separation ability.[69]

These zeolite membrane compositions that have been employed to separate CO_2/CH_4 mixtures are essentially the same composition employed for CO_2/N_2 and include zeolites Y,[87] zeolite T,[83] ZSM-5,[85] SAPO-34,[101–106] DD3R[86] and AlPO-18.[82] Since the size difference between CO_2 and CH_4 is larger (as compared to CO_2-N_2), these zeolite membranes have shown high CO_2/CH_4 selectivities due to a combination of differences in diffusivities and preferential CO_2 adsorption. However, most of these zeolite compositions have shown only moderate CO_2 permeability. SAPO-34 has shown the best separation performance of any zeolite membrane composition, even at high feed pressures.[101–103]

Several MOF compositions have been successfully prepared in membrane form, and have been evaluated for CO_2/CH_4 gas mixtures. These compositions include: ZIF-8,[9,31,53,55,62,90] HKUST-1,[63,95] MOF-5,[36] MIL-53,[59] $[Cu_2L_2P]n$,[34] SIM-1,[42] ZIF-7,[56] ZIF-22,[91] ZIF-69,[52] ZIF-95[107] and ZIF-90.[28,47,92,93] Some of these MOFs have displayed low to moderate CO_2/CH_4 selectivities with moderate to high CO_2 permeances.

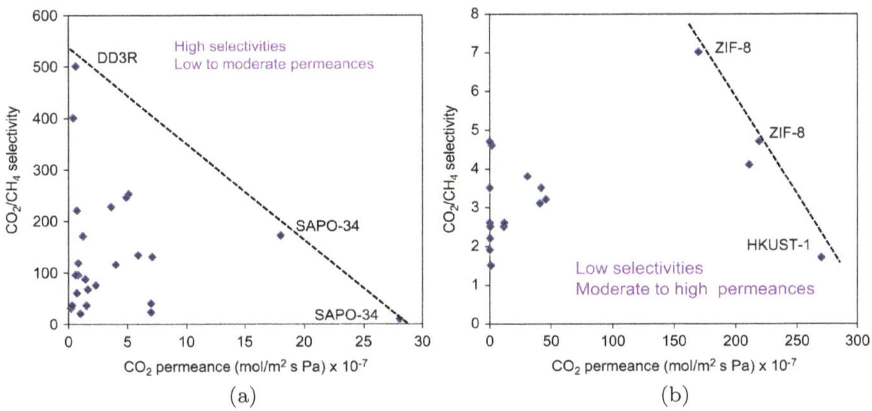

Figure 6.6. CO_2/CH_4 separation selectivity vs CO_2 permeance for (a) zeolite and (b) MOF membranes. (Reprinted from Ref. 24 with permission from Elsevier.)

The flexibility of the MOF framework is in part responsible for the low observed separation selectivities.

Figure 6.6 summarizes the performance of (a) zeolite membranes and (b) MOF membranes for CO_2/CH_4 separation. Similar to CO_2/N_2 separations, zeolite membranes display much higher CO_2/CH_4 separation selectivities as compared to MOF membranes. The enhanced separation selectivities on zeolite membranes have been associated to molecular sieving effect within zeolite pores. On the other hand, the limited selectivities observed in MOFs has been related to the presence of organic linkers which lead to a flexible MOF structure (non-rigid pore size) which may limit a potential molecular sieving effect. In terms of CO_2 permeances, MOFs display typically higher values as compared to zeolites due to the framework flexibility and higher surface areas. In Figure 6.6, we have included the zeolite and MOF compositions that lie in the upper bound limit. For example in Figure 6.6a, zeolite DD3R shows the highest CO_2/CH_4 separation selectivity but low CO_2 permeance.[86] DD3R is a microporous structure consisting of cages, each connected by three windows of 8-membered rings forming a 2-dimensional pore structure with pore sizes of 0.36×0.44 nm. The authors suggested that the high separation selectivities were related to steric effects introduced by the window opening of DD3R leading to molecular sieving and activated transport, as well as to preferential adsorption effects observed for CO_2/CH_4 mixtures.[86] The other zeolite present on the upper bound limit is SAPO-34 Figure 6.6a. SAPO-34 membranes have shown the best overall separation performance for CO_2/CH_4 mixtures (high selectivities with moderate to high CO_2 permeances[101–106] as shown in Figure 6.6a. There is a general agreement that SAPO-34 membranes display high CO_2/CH_4 selectivities due to a combination of differences in diffusivity and competitive adsorption, and to a sharp molecular sieving effect. The moderate to high CO_2 permeances have been achieved by preparing SAPO-34 thin membranes.[101–106]

In the case of MOF compositions, as shown in Figure 6.6b, ZIF-8, a zeolitic imidazolate framework with limiting pore aperture of 0.34 nm shows the highest CO_2/CH_4 separation selectivity, and high CO_2 permeance.[53] The high observed CO_2 permeances on ZIF-8 membranes were associated not only due to the presence of small crystals with a narrow size distribution, which led to thin membranes, but also due to the textural properties of the alumina porous support (where the membrane was grown). Although ZIF-8 is composed of large 1.16 nm pores and small pore apertures of 0.34 nm, density functional theory simulation data suggest that the smaller pores are the preferential adsorption sites for CO_2 molecules. Therefore, the small pore aperture of ZIF-8 favored the diffusion of CO_2 over CH_4 resulting in moderate separation selectivities. HKUST-1 showed the highest CO_2 permeance[95] and low separation selectivities, which were associated to the large pore size of MOF structure. Figure 6.7 shows representative SEM images ZIF-8 membranes that have shown the best separation performance for CO_2/CH_4 gas mixtures.

Figure 6.7. ZIF-8 membrane (a) top view and (b) cross section view employed to separate CO_2/CH_4 gas mixtures. (Reprinted from Ref. 55 with permission from American Chemical Society.)

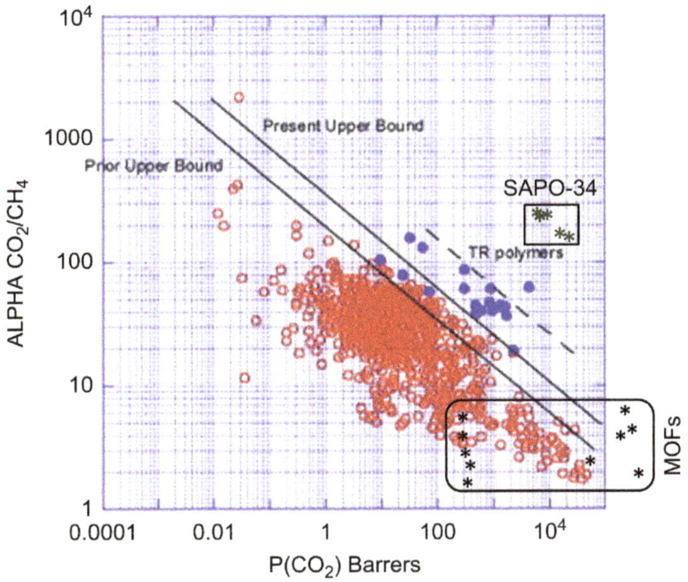

Figure 6.8. Robeson plot for CO_2/CH_4 gas mixtures. Red and blue points represent experimental points for polymeric membranes. For comparison SAPO-34 zeolite membranes are included. CO_2 permeabilites for the four points shown out of the graph are $>10^6$ Barrers. (Reprinted from Ref. 24 with permission from Elsevier.)

Figure 6.8 shows the Robeson plot for CO_2/CH_4 gas mixtures. For comparison, the performance of SAPO-34 membranes has been included. The experimental data points for SAPO-34 membranes are significantly above the upper bound, indicating that these zeolite membranes perform better than any polymeric and MOF membrane. Although the separation performance of SAPO-34 membranes may be enough for natural gas treatment, as pure inorganic membrane, SAPO-34 membrane cost (membrane module) can limit its potential application at industrial level. Current efforts in decreasing the cost of SAPO-34 membranes have been documented.[102,103] For most MOF membranes, the separation performance lies below the upper bound. Only ZIF-8 membranes perform better than conventional polymeric membranes and comparable to thermally rearranged polymeric membranes.

6.6. MOF membranes for CO_2/H_2 separations

Membranes represent an alternative technology for carbon dioxide capture and hydrogen purification at hydrogen production facilities and Integrated

Gasification Combined Cycle (IGCC) power plants due to their simple operation mode and potentially low energy consumption.[108] Hydrogen selective membranes[73,109–113] and CO_2-selective membranes [114–118] have been employed. All these membranes are composed of polymeric materials or dense metal phases. Only few examples, discussed below, on porous crystalline membranes for CO_2/H_2 (CO_2-selective) separation have been reported. Carbon dioxide selective membranes have attracted significant interest because pure hydrogen can be retained at high pressure, thus avoiding high recompression costs. In particular, these membranes are of great relevance for CO_2/H_2 separations in steam methane reforming where it is desirable to have pure hydrogen at high pressure by permeating CO_2 from syngas,[114–118] as well as in air-blown gasification processes, where CO_2 needs to be removed from syngas containing H_2 and N_2.[114]

MOF membranes have shown to be only hydrogen selective rather than CO_2 selective. Representative examples of these H_2 selective MOF membranes include: ZIF-7,[57,66] ZIF-22,[91] ZIF-90,[91,93] and ZIF-8.[90] These membranes have shown H_2 permeances in the 9×10^{-9} to 2.5×10^{-7} mol/m^{2}s Pa range with H_2/CO_2 separation selectivities in the 6–14 range. A review on porous inorganic membranes for CO_2 capture has been reported.[119] In this review, progress in the synthesis of porous inorganic membranes and their potentials for CO_2 capture in pre- and postcombustion scenarios is discussed.

6.7. Concluding remarks and outlook

MOF membranes for CO_2 separation is a research field that has evolved rapidly. The high porosity and high surface areas (among other MOF properties) of several MOF membranes has led to high CO_2 permeances. Another unique feature of MOF membranes is related to its unprecedented high CO_2 uptakes which in principle should translate into high CO_2 ideal selectivities over other gases. As compared to other microporous crystalline structures such as zeolites, MOFs (if prepared in membrane form) offer a wide range of pore sizes leading to infinite number of structures with flexible pore sizes and structures. Finally, due to the presence of organic linkers in MOF structures, proper functionalization of MOFs may result in pore size control as well as adsorption properties control. This in principle would allow one to rationally design "tailored MOF membranes" for CO_2 separation.

Although MOF membranes are very promising for CO_2 separation, we need to consider practical challenges that may impede the future application of these membranes for CO_2 related gas separations. First of all, it is important to note that even if MOF crystals exhibiting unique and remarkable separation properties is prepared in powder or particle form, the same material may be not suitable for membrane preparation due to limited adhesion of the membrane to the support, considerable adsorption induced stresses, poor crystal intergrowth, large crystal size leading to non-continuous membranes, poor reproducibility, etc. In addition, the potential "sharp molecular sieving" for MOFs in CO_2 separations, still needs to be demonstrated. The flexible nature of MOF structures may limit their potential sharp molecular sieving effect.

MOF membrane synthesis approaches aim to improve membrane intergrowth, control thickness, and reduce concentration of "non-selective pores" may result in high quality membranes able to effectively separate CO_2. These synthetic approaches should mainly consider (a) the preparation of microporous crystals with small size and narrow size distribution that could allow for better "crystal packing" during membrane growth; (b) chemical, physical, thermal or other seeding techniques to improve membrane intergrowth and (c) effective membrane-support interaction. In addition, novel synthesis methods for the preparation of MOF membranes could lead to improved membranes displaying superior separation performance. These novel methods (among others) may include electrochemical synthesis[30,120] liquid phase deposition,[34] contra-diffusion method [35] and evaporation induced self-assembly approaches.[121–123] These are potential synthetic approaches that can lead to MOF membranes displaying controlled thickness, crystal size, and improved intergrowth. MOF membrane processing challenges can be circumvented by dispersing the MOFs in polymers and can achieve high gas separation performances above the Robeson upper bound compared to the pure polymer membranes. Finally, some important challenges to overcome for considering MOF membranes for practical applications mainly include poor reproducibility; demonstration of long term and chemical stability, high cost, and limited separation selectivities. We discuss these challenges in the last chapter of this book.

References

1. DOE/EIA. EIA estimates on Global Warming Potentials: Intergovernmental Panel on Climate Changes, Climate Change: the physical Science Basis, Cambridge UK, Cambridge University Press (2007).
2. DOE/EIA. *Annu. Energy Outlook* **0383**, (2008).
3. Baker, R. W. Future directions of membrane gas separation technology. *Ind. Eng. Chem. Res.* **41**, 1393–1411 (2002).
4. Lin, H., Van Wagner, E., Raharjo, R., Freeman, B. D. & Roman, I. High-performance polymer membranes for natural-gas sweetening. *Adv. Mater.* **18**, 39–44 (2006).
5. Alexander Stern, S. Polymers for gas separations: The next decade. *J. Memb. Sci.* **94**, 1–65 (1994).
6. Millward, A. R. & Yaghi, O. M. Metal-organic frameworks with exceptionally high capacity for storage of carbon dioxide at room temperature. *J. Am. Chem. Soc.* **127**, 17998–17999 (2005).
7. Bourrelly, S. Llewellyn, P. L., Serre, C., Millange, F., Loiseau, T. & Férey, G., Different adsorption behaviors of methane and carbon dioxide in the isotypic nanoporous metal terephthalates MIL-53 and MIL-47. *J. Am. Chem. Soc.* **127**, 13519–13521 (2005).
8. Llewellyn, P. L., Bourrelly, S., Serre, C., Vimont, A., Daturi, M., Hamon, L., De Weireld, G., Chang, J-S., Hong, D-Y., Kyu Hwang, Y., Hwa Jhung, S. & Férey, G., High uptakes of CO_2 and CH_4 in mesoporous metal-organic frameworks MIL-100 and MIL-101. Langmuir **24**, 7245–7250 (2008).
9. Venna, S. R., Zhu, M., Li, S. & Carreon, M. A. Knudsen diffusion through ZIF-8 membranes synthesized by secondary seeded growth, *J. Porous Mater.* **21**, 235–240 (2013).
10. Venna, S. R. & Carreon, M. A. Synthesis of SAPO-34 crystals in the presence of crystal growth inhibitors. *J. Phys. Chem. B* **112**, 16261–16265 (2008).
11. Maghsoudi, H., Soltanieh, M., Bozorgzadeh, H. & Mohamadalizadeh, A. Adsorption isotherms and ideal selectivities of hydrogen sulfide and carbon dioxide over methane for the Si-CHA zeolite: Comparison of carbon dioxide and methane adsorption with the all-silica DD3R zeolite. *Adsorption* **19**, 1045–1053 (2013).
12. Calleja, G., Pau, J. & Calles, J. A. Pure and multicomponent adsorption equilibrium of carbon dioxide, ethylene, and propane on ZSM-5 zeolites with different Si/Al ratios. *J. Chem. Eng. Data* **43**, 994–1003 (1998).
13. Dunne, J. & Myers, A. L. Adsorption of gas mixtures in micropores: Effect of difference in size of adsorbate molecules. *Chem. Eng. Sci.* **49**, 2941–2951 (1994).
14. Pakseresht, S. Equilibrium isotherms for CO, CO_2, CH_4 and C_2H_4 on the 5A molecular sieve by a simple volumetric apparatus. *Sep. Purif. Technol.* **28**, 53–60 (2002).
15. Park, K. S., Ni, Z., Côté, A.P., Choi, J. Y., Huang, R., Uribe-Romo, F. J., Chae, H. K., O'Keeffe, M. & Yaghi, O. M., Exceptional chemical and thermal stability of zeolitic imidazolate frameworks. *Proc. Natl. Acad. Sci. U. S. A.* **103**, 10186–10191 (2006).
16. Hayashi, H., Côté, A. P., Furukawa, H., O'Keeffe, M. & Yaghi, O. M. Zeolite A imidazolate frameworks, *Nat. Mater.* **6**, 501–506 (2007).
17. An, J., Fiorella, R. P., Geib, S. J. & Rosi, N. L. Synthesis, structure, assembly, and modulation of the CO_2 adsorption properties of a zinc-adeninate macrocycle. *J. Am. Chem. Soc.* **131**, 8401–8403 (2009).

18. An, J., Geib, S. J. & Rosi, N. L. Cation-triggered drug release from a porous zinc-adeninate metal-organic framework. *J. Am. Chem. Soc.* **131**, 8376–8377 (2009).
19. An, J., Geib, S. J. & Rosi, N. L. High and selective CO_2 uptake in a cobalt adeninate metal-organic framework exhibiting pyrimidine- and amino-decorated pores. *J. Am. Chem. Soc.* **132**, 38–39 (2010).
20. An, J. & Rosi, N. L. Tuning MOF CO_2 adsorption properties via cation exchange. *J. Am. Chem. Soc.* **132**, 5578–5579 (2010).
21. An, J., Shade, C. M., Chengelis-Czegan, D. A., Petoud, S. & Rosi, N. L. Zinc-adeninate metal-organic framework for aqueous encapsulation and sensitization of near-infrared and visible emitting lanthanide cations. *J. Am. Chem. Soc.* **133**, 1220–1223 (2011).
22. Li, T., Chen, D. L. Sullivan, J. E., Kozlowski, M. T., Johnson, J. K. & Rosi, N. L., Systematic modulation and enhancement of $CO_2 : N_2$ selectivity and water stability in an isoreticular series of bio-MOF-11 analogues. *Chem. Sci.* **4**, 1746–1755 (2013).
23. Li, J.-R., Ma. Y., McCarthy, M. C., Sculley, J., Yu, J., Jeong, H. K., Balbuene, P. B. & Zhou, H. C., Carbon dioxide capture-related gas adsorption and separation in metal-organic frameworks. *Coord. Chem. Rev.* **255**, 1791–1823 (2011).
24. Venna, S.R. & Carreon, M.A. Metal-organic framework membranes for carbon dioxide separation, *Chemical Engineering Science* **124**, 3–19 (2015).
25. Gascon, J. & Kapteijn, F. Metal-organic framework membranes — high potential, bright future? *Angew. Chem. Int. Ed. Engl.* **49,** 1530–1532 (2010).
26. Nagaraju, D., Bhagat, D. G., Banerjee, R. & Kharul, U. K. *In situ* growth of metal-organic frameworks on a porous ultrafiltration membrane for gas separation. *J. Mater. Chem. A* **1**, 8828 (2013).
27. Kwon, H. T. & Jeong, H.-K. *In situ* synthesis of thin zeolitic-imidazolate framework ZIF-8 membranes exhibiting exceptionally high propylene/propane separation. *J. Am. Chem. Soc.* **135**, 10763–10768 (2013).
28. Huang, A., Liu, Q., Wang, N. & Caro, J. Organosilica functionalized zeolitic imidazolate framework ZIF-90 membrane for CO_2/CH_4 separation. *Microporous Mesoporous Mater.* **192**, 18–22 (2014).
29. Xie, Z. Yand, J., Wand, J., Bai, J., Yin, H., Yuan, B., Lu, J., Zhang, Y., Zhou, L. & Duan, C., Deposition of chemically modified α-Al_2O_3 particles for high performance ZIF-8 membrane on a macroporous tube. *Chem. Commun. (Camb).* **48**, 5977–5979 (2012).
30. Li, M. & Dincă, M. Selective formation of biphasic thin films of metal-organic frameworks by potential-controlled cathodic electrodeposition. *Chem. Sci.* **5**, 107 (2014).
31. Bux, H. Liang, F. Li, Y., Cravillon, J., Wiebcke, M. & Caro, J., Zeolitic imidazolate framework membrane with molecular sieving properties by microwave-assisted solvothermal synthesis. *J. Am. Chem. Soc.***131**, 16000–16001 (2009).
32. Yoo, Y., Lai, Z. & Jeong, H.-K. Fabrication of MOF-5 membranes using microwave-induced rapid seeding and solvothermal secondary growth. *Microporous Mesoporous Mater.* **123**, 100–106 (2009).
33. Shah, M., McCarthy, M. C., Sachdeva, S., Lee, A. K. & Jeong, H.-K. Current status of metal-organic framework membranes for gas separations: Promises and challenges. *Ind. Eng. Chem. Res.* **51**, 2179–2199 (2012).
34. Bétard, A., Bux, H., Henke, S., Zacher, D., Caro, J. & Fischer, A. A., Fabrication of a CO_2-selective membrane by stepwise liquid-phase deposition of an alkylether

functionalized pillared-layered metal-organic framework $[Cu_2L_2P]_n$ on a macroporous support. *Microporous Mesoporous Mater.* **150**, 76–82 (2012).
35. Yao, J., Dong, D., Li, D., He, L., Xu, G., Wang, H., *et al.* Contra-diffusion synthesis of ZIF-8 films on a polymer substrate. *Chem. Commun. (Camb).* **47**, 2559–2561 (2011).
36. Liu, Y., Ng, Z., Khan, E. A., Jeong, H. K., Ching, C. & Lai, Z., Synthesis of continuous MOF-5 membranes on porous α-alumina substrates. *Microporous Mesoporous Mater.* **118**, 296–301 (2009).
37. Liu, Y., Hu, E., Khan, E. A. & Lai, Z. Synthesis and characterization of ZIF-69 membranes and separation for CO_2/CO mixture. *J. Memb. Sci.* **353**, 36–40 (2010).
38. Huang, A. & Caro, J. Covalent post-functionalization of zeolitic imidazolate framework ZIF-90 membrane for enhanced hydrogen selectivity. *Angew. Chem. Int. Ed. Engl.* **50**, 4979–4982 (2011).
39. Huang, A., Dou, W. & Caro, J. Steam-stable zeolitic imidazolate framework ZIF-90 membrane with hydrogen selectivity through covalent functionalization. *J. Am. Chem. Soc.* **132**, 15562–15564 (2010).
40. Shah, M., Kwon, H. T., Tran, V., Sachdeva, S. & Jeong, H.-K. One step in situ synthesis of supported zeolitic imidazolate framework ZIF-8 membranes: Role of sodium formate. *Microporous Mesoporous Mater.* **165**, 63–69 (2013).
41. Cacho-Bailo, F., Seoane, B., Téllez, C. & Coronas, J. ZIF-8 continuous membrane on porous polysulfone for hydrogen separation, *J. Memb. Sci.* **464**, 119-126 (2014).
42. Aguado, S., Nicolas, C-H., Moizan-Basle, V., Nieto, C., Amrouche, H., Bats, N., Auderbarnad, N. & Ferrusseng, D., Facile synthesis of an ultramicroporous MOF tubular membrane with selectivity towards CO_2. *New J. Chem.* **35**, 41 (2011).
43. Xu, G., Yao, J., Wang, K., He, L., Webley, P. A., Chen, C-S. & Wang, H., Preparation of ZIF-8 membranes supported on ceramic hollow fibers from a concentrated synthesis gel. *J. Memb. Sci.* **385–386**, 187–193 (2011).
44. Pan, Y., Wang, B. & Lai, Z. Synthesis of ceramic hollow fiber supported zeolitic imidazolate framework-8 (ZIF-8) membranes with high hydrogen permeability. *J. Memb. Sci.* **421–422**, 292–298 (2012).
45. Huang, K., Dong, Z., Li, Q. & Jin, W. Growth of a ZIF-8 membrane on the inner-surface of a ceramic hollow fiber via cycling precursors. *Chem. Commun. (Camb).* **49**, 10326–10328 (2013).
46. Mao, Y., Li, J., Gao, W., Ying, Y., Sun, L. & Peng, X. Pressure-assisted synthesis of HKUST-1 thin film on polymer hollow fiber at room temperature toward gas separation. *ACS Appl. Mater. Interfaces* **6**, 4473–4479 (2014).
47. Brown, A. J., Johnson, J. R., Lydon, M. E., Koros, W. J., Jones, C.W. & Nair, S., Continuous polycrystalline zeolitic imidazolate framework-90 membranes on polymeric hollow fibers. *Angew. Chem. Int. Ed. Engl.* **51**, 10615–10618 (2012).
48. Brown, A. J. Brunelli, N. A., Eum, K., Rashidi, F., Johnson, J. R., Koros, W. J., Jones, C. W. & Nair, S., Separation membranes. Interfacial microfluidic processing of metal-organic framework hollow fiber membranes. *Science* **345**, 72–75 (2014).
49. Guerrero, V. V., Yoo, Y., McCarthy, M. C. & Jeong, H.-K. HKUST-1 membranes on porous supports using secondary growth. *J. Mater. Chem.* **20**, 3938 (2010).
50. Nan, J., Dong, X., Wang, W. & Jin, W. Formation mechanism of metal-organic framework membranes derived from reactive seeding approach. *Microporous Mesoporous Mater.* **155**, 90–98 (2012).

51. Ranjan, R. & Tsapatsis, M. Microporous metal-organic framework membrane on porous support using the seeded growth method. *Chem. Mater.* **21**, 4920–4924 (2009).
52. Liu, Y., Zeng, G., Pan, Y. & Lai, Z. Synthesis of highly c-oriented ZIF-69 membranes by secondary growth and their gas permeation properties. *J. Memb. Sci.* **379**, 46–51 (2011).
53. Venna, S. R. & Carreon, M. A. Highly permeable zeolite imidazolate framework-8 membranes for CO_2/CH_4 separation. *J. Am. Chem. Soc.* **132**, 76–8 (2010).
54. Zhang, F., Zou, X, Gao, X., Fan, S., Sun, F., Ren, H. & Zhu, G., Hydrogen Selective NH_2-MIL-53(Al) MOF Membranes with High Permeability. *Adv. Funct. Mater.* **22**, 3583–3590 (2012).
55. McCarthy, M. C., Varela-Guerrero, V., Barnett, G. V. & Jeong, H.-K. Synthesis of zeolitic imidazolate framework films and membranes with controlled microstructures. *Langmuir* **26**, 14636–14641 (2010).
56. Li, Y.-S., Liang, F. Y., Bux, H., Feldhoff, A., Yand, W. S. & Caro. J., Molecular sieve membrane: supported metal-organic framework with high hydrogen selectivity. *Angew. Chem. Int. Ed. Engl.* **49**, 548–551 (2010).
57. Li, Y., Liang, F., Bux, H., Yang, W. & Caro, J. Zeolitic imidazolate framework ZIF-7 based molecular sieve membrane for hydrogen separation. *J. Memb. Sci.* **354**, 48–54 (2010).
58. Zhou, S., Zou, X., Sun, F., Zhang, F., Fan , S., Zhao, H., Schiestel, T. & Zhu, G., Challenging fabrication of hollow ceramic fiber supported $Cu_3(BTC)_2$ membrane for hydrogen separation. *J. Mater. Chem.* **22**, 10322 (2012).
59. Hu, Y., Dong, X., Nan, J., Jin, W., Ren, X., Xu, N. & Lee, Y.M., Metal-organic framework membranes fabricated via reactive seeding. *Chem. Commun. (Camb).* **47**, 737–739 (2011).
60. Dong, X. & Lin, Y. S. Synthesis of an organophilic ZIF-71 membrane for pervaporation solvent separation. *Chem. Commun. (Camb).* **49**, 1196–1198 (2013).
61. Wang, W., Dong, X., Nan, J, Jin, W., Hu, Z., Chen, Y. & Jiand, J., A homochiral metal-organic framework membrane for enantioselective separation. *Chem. Commun. (Camb).* **48**, 7022–7024 (2012).
62. Pan, Y. & Lai, Z. Sharp separation of C_2/C_3 hydrocarbon mixtures by zeolitic imidazolate framework-8 (ZIF-8) membranes synthesized in aqueous solutions. *Chem. Commun. (Camb).* **47**, 10275–10277 (2011).
63. Guo, H., Zhu, G., Hewitt, I. J. & Qiu, S. "Twin copper source" growth of metal-organic framework membrane: Cu(3)(BTC)(2) with high permeability and selectivity for recycling H(2). *J. Am. Chem. Soc.* **131**, 1646–1647 (2009).
64. Zhang, X., Liu, Y., Kong, L., Liu, H., Qiu, J., Han, H., Weng, L-T, Yeung, K.L. & Zhu, E., A simple and scalable method for preparing low-defect ZIF-8 tubular membranes. *J. Mater. Chem. A* **1**, 10635–10638 (2013).
65. Abdollahian, Y., Hauser, J. L., Colinas, I.R., Agustin, C., Ichimura, A. S. & Oliver, S. R. IRMOF thin films templated by oriented zinc oxide nanowires. *Cryst. Growth Des.* **14**, 1506–1509 (2014).
66. Li, Y.-S., Bux, H., Flendhot, A., Li, G-L., Yang, W-S. & Caro, J., Controllable synthesis of metal-organic frameworks: From MOF nanorods to oriented MOF membranes. *Adv. Mater.* **22**, 3322–3326 (2010).

67. Hara, N., Yoshimune, M., Negishi, H., Haraya, K., Hara, S. & Yamaguchi, T., Metal-organic framework membranes with layered structure prepared within the porous support. *RSC Adv.* **3**, 14233–14236 (2013).
68. Favre, E. Carbon dioxide recovery from post-combustion processes: Can gas permeation membranes compete with absorption? *J. Memb. Sci.* **294**, 50–59 (2007).
69. Koros, W. J. & Mahajan, R. Pushing the limits on possibilities for large scale gas separation: which strategies? *J. Memb. Sci.* **181**, 141 (2001).
70. Lin, H. & Freeman, B. D. Gas solubility, diffusivity and permeability in poly(ethylene oxide). *J. Memb. Sci.* **239**, 105–117 (2004).
71. Okamoto, K.-I.; Umeo, N.; Okamyo, S.; Tanaka, K. & Kita, H. Selective permeation of carbon dioxide over nitrogen through polyethylene oxide containing polyimide membranes. *Chem. Lett.* **22**, 225–228 (1993).
72. Robeson, L. M. The upper bound revisited. *J. Memb. Sci.* **320**, 390–400 (2008).
73. Bredesen, R., Jordal, K. & Bolland, O. High-temperature membranes in power generation with CO_2 capture. *Chem. Eng. Process. Process Intensif.* **43**, 1129–1158 (2004).
74. Huang, J., Zou, J. & Ho, W. S. W. Carbon dioxide capture using a CO_2-selective facilitated transport membrane. *Ind. Eng. Chem. Res.* **47**, 1261–1267 (2008).
75. Zolandz, R.R. & Fleming, G. K. Definitions. In *Membrane Handbook*; Ho, W.S.W.; Sirkar, K.K., Eds.; Van Nostrand Reinhold, Inc.: New York, 19–24 (1992).
76. Thompson, R. W. Recent advances in the understanding of zeolite synthesis, molecular sievesm. H.G. Karge, J. Weitkamp, (Eds.) Springer, Berlin Heidelberg (1998).
77. Aoki, K., Kusakabe, K. & Morooka, S. Separation of gases with an A-Type zeolite membrane. *Ind. Eng. Chem. Res.* **39**, 2245–2251 (2000).
78. Baertsch, C. D., Funke, H. H., Falconer, J. L. & Noble, R. D. Permeation of aromatic hydrocarbon vapors through silicalite-zeolite membranes. *J. Phys. Chem.* **100**, 7676–7679 (1996).
79. Bowen, T. C., Noble, R. D. & Falconer, J. L. Fundamentals and applications of pervaporation through zeolite membranes. *J. Memb. Sci.* **245**, 1–33 (2004).
80. Hong, M., Li, S., Falconer, J. L. & Noble, R. D. Hydrogen purification using a SAPO-34 membrane. *J. Memb. Sci.* **307**, 277–283 (2008).
81. Snyder, M. A. & Tsapatsis, M. Hierarchical nanomanufacturing: from shaped zeolite nanoparticles to high-performance separation membranes. *Angew. Chem. Int. Ed. Engl.* **46**, 7560–73 (2007).
82. Carreon, M. L., Li, S. & Carreon, M. A. AlPO-18 membranes for CO_2/CH_4 separation. *Chem. Commun. (Camb).* **48**, 2310–2 (2012).
83. Cui, Y., Kita, H. & Okamoto, K. Preparation and gas separation performance of zeolite T membrane. *J. Mater. Chem.* **14**, 924 (2004).
84. Li, S. & Fan, C. Q. High-Flux SAPO-34 Membrane for CO_2/N_2 Separation. *Ind. Eng. Chem. Res.* **49**, 4399–4404 (2010).
85. Shin, D. W., Hyun, S. H., Cho, C. H. & Han, M. H. Synthesis and CO_2/N_2 gas permeation characteristics of ZSM-5 zeolite membranes. *Microporous Mesoporous Mater.* **85**, 313–323 (2005).
86. Van den Bergh, J., Zhu, W., Gascon, J., Moulijn, J. A. & Kapteijn, F. Separation and permeation characteristics of a DD3R zeolite membrane. *J. Memb. Sci.* **316**, 35–45 (2008).

87. White, J. C., Dutta, P. K., Shqau, K. & Verweij, H. Synthesis of ultrathin zeolite Y membranes and their application for separation of carbon dioxide and nitrogen gases. *Langmuir* **26**, 10287–10293 (2010).
88. Huang, X.-C., Lin, Y.-Y., Zhang, J.-P. & Chen, X.-M. Ligand-directed strategy for zeolite-type metal-organic frameworks: zinc(II) imidazolates with unusual zeolitic topologies. *Angew. Chem. Int. Ed. Engl.* **45**, 1557–1559 (2006).
89. Yaghi, O. M., O'Keeffe, M., Ockwig, N. W., Chae, H. K., Eddaoudi, M. & Kim, J., Reticular synthesis and the design of new materials. *Nature* **423,** 705–714 (2003).
90. Bux, H., Feldhoff, A., Cravillon, J., Wiebcke, M., Li, Y.S. & Caro, J., Oriented zeolitic imidazolate framework-8 membrane with sharp H_2/C_3H_8 molecular sieve separation. *Chem. Mater.* **23**, 2262–2269 (2011).
91. Huang, A., Bux, H., Steinbach, F. & Caro, J. Molecular-sieve membrane with hydrogen permselectivity: ZIF-22 in LTA topology prepared with 3-aminopropyltriethoxysilane as covalent linker. *Angew. Chem. Int. Ed. Engl.* **49**, 4958–4961 (2010).
92. Huang, A. & Caro, J. Covalent post-functionalization of zeolitic imidazolate framework ZIF-90 membrane for enhanced hydrogen selectivity. *Angew. Chem. Int. Ed. Engl.* **50**, 4979–4982 (2011).
93. Huang, A., Dou, W. & Caro, J. Steam-stable zeolitic imidazolate framework ZIF-90 membrane with hydrogen selectivity through covalent functionalization. *J. Am. Chem. Soc.* **132**, 15562–15564 (2010).
94. Zhao, Z., Ma, X., Kasik, A., Li, Z. & Lin, Y. S. Gas separation properties of metal-organic framework (MOF-5) membranes. *Ind. Eng. Chem. Res.* **52**, 1102–1108 (2013).
95. Guerrero, V. V., Yoo, Y., McCarthy, M. C. & Jeong, H.-K. HKUST-1 membranes on porous supports using secondary growth. *J. Mater. Chem.* **20**, 3938 (2010).
96. Merkel, T. C., Lin, H., Wei, X. & Baker, R. Power plant post-combustion carbon dioxide capture: An opportunity for membranes. *J. Memb. Sci.* **359**, 126–139 (2010).
97. Baker, R. W. & Lokhandwala, K. Natural gas processing with membranes: An overview. *Ind. Eng. Chem. Res.* **47**, 2109–2121 (2008).
98. Spillman. R.W. Economics of gas separation by membranes. *Chem. Eng. Prog.* **85**, 41 (1989).
99. Ohya, H. & V V Kudryavsev, S. I. S. Polyimide Membranes: Applications, Fabrications and Properties, CRC Press (1997).
100. Baker, R.W., Pinnau, I.; He, Z.; Amo, K.D.; DaCosta, A.R. & Daniels, R. . Nitrogen gas separation using organic vapor resistant membranes (2003).
101. Funke, H. H., Chen, M. Z., Prakash, A. N., Falconer, J. L. & Noble, R. D. Separating molecules by size in SAPO-34 membranes. *J. Memb. Sci.* **456**, 185–191 (2014).
102. Ping, E. W., Zhou, R., Funke, H. H., Falconer, J. L. & Noble, R. D. Seeded-gel synthesis of SAPO-34 single channel and monolith membranes, for CO_2/CH_4 separations. *J. Memb. Sci.* **415-416**, 770–775 (2012).
103. Zhou, R., Ping, E. W., Funke, H. H., Falconer, J. L. & Noble, R. D. Improving SAPO-34 membrane synthesis. *J. Memb. Sci.* **444**, 384–393 (2013).
104. Carreon, M. A., Li, S., Falconer, J. L. & Noble, R. D. SAPO-34 Seeds and membranes prepared using multiple structure directing agents. *Adv. Mater.* **20**, 729–732 (2008).
105. Carreon, M. A., Li, S., Falconer, J. L. & Noble, R. D. Alumina-supported SAPO-34 membranes for CO_2/CH_4 separation. *J. Am. Chem. Soc.* **130**, 5412–5413 (2008).

106. Venna, S. R. & Carreon, M. A. Amino-Functionalized SAPO-34 Membranes for CO_2/CH_4 and CO_2/N_2 Separation. *Langmuir* **27**, 2888–2894 (2011).
107. Huang, A., Chen , Y., Wang, N., Hu, Z., Jiang, J. & Caro, J., A highly permeable and selective zeolitic imidazolate framework ZIF-95 membrane for H_2/CO_2 separation. *Chem. Commun. (Camb).* **48**, 10981–10983 (2012).
108. Merkel, T. C., Zhou, M. & Baker, R. W. Carbon dioxide capture with membranes at an IGCC power plant. *J. Memb. Sci.* **389**, 441–450 (2012).
109. Ramasubramanian, K., Zhao, Y. & Winston Ho, W. S. CO_2 capture and H_2 purification: Prospects for CO_2 -selective membrane processes. *AIChE J.* **59**, 1033–1045 (2013).
110. Liu, P. K., Sahimi, M. & Tsotsis, T. T. Process intensification in hydrogen production from coal and biomass via the use of membrane-based reactive separations. *Curr. Opin. Chem. Eng.* **1**, 342–351 (2012).
111. Guazzone, F., Catalano, J., Mardilovich, I. P., Kniep, J., Pande, S., Wu, T., Lambrecht, R. C., Datta, S., Kazantzis, N. K. & Ma, Y. H., Gas permeation field tests of composite Pd and Pd–Au membranes in actual coal derived syngas atmosphere. *Int. J. Hydrogen Energy* **37**, 14557–14568 (2012).
112. Ku, A. Y., Kulkarni, P., Shisler, R. & Wei, W. Membrane performance requirements for carbon dioxide capture using hydrogen-selective membranes in integrated gasification combined cycle (IGCC) power plants. *J. Memb. Sci.* **367**, 233–239 (2011).
113. Ma, Y.H. Hydrogen separation membranes, Li, N.N. Fane, A.G. Ho, W.S. Matsuura T. (Eds.), Advanced Membrane Technology and Applications, John Wiley & Sons, Hoboken, New Jersey (2008), pp. 671–684.
114. Lin, H., He, Z., Sun, Z., Vu, J., Ng, A. Mohamood, M., Kniep, J., Merkel, T. C., Wu, T. & Lambrecht, R.C., CO_2-selective membranes for hydrogen production and CO_2 capture – Part I: Membrane development. *J. Memb. Sci.* **457**, 149–161 (2014).
115. Reijerkerk, S. R., Nijmeijer, K., Ribeiro, C. P., Freeman, B. D. & Wessling, M. On the effects of plasticization in CO_2/light gas separation using polymeric solubility selective membranes. *J. Memb. Sci.* **367**, 33–44 (2011).
116. Yave, W., Car, A., Wind, J. & Peinemann, K.-V. Nanometric thin film membranes manufactured on square meter scale: ultra-thin films for CO_2 capture. *Nanotechnology* **21**, 395301 (2010).
117. Lin, H., Van Wagner, E., Freeman, B. D., Toy, L. G. & Gupta, R. P. Plasticization-enhanced hydrogen purification using polymeric membranes. *Science* **311**, 639–642 (2006).
118. Bondar, V. I., Freeman, B. D. & Pinnau, I. Gas transport properties of poly(ether-b-amide) segmented block copolymers. *J. Polym. Sci. Part B Polym. Phys.* **38**, 2051–2062 (2000).
119. Pera-Titus, M. Porous inorganic membranes for CO_2 capture: present and prospects. *Chem. Rev.* **114**, 1413–1492 (2014).
120. Wade, C. R., Li, M. & Dincă, M. Facile deposition of multicolored electrochromic metal-organic framework thin films. *Angew. Chem. Int. Ed. Engl.* **52**, 13377–13381 (2013).
121. Brinker, C. J., Lu, Y., Sellinger, A. & Fan, H. Evaporation-induced self-assembly: nanostructures made easy. *Adv. Mater.* **11**, 579–585 (1999).

122. Grosso, D. *et al.* Fundamentals of mesostructuring through evaporation-induced self-assembly. *Adv. Funct. Mater.* **14**, 309–322 (2004).
123. Carreon M.A. & Guliants V.V. "Mesostructuring Metal Oxides through Evaporation Induced Self-Assembly: Fundamentals and Applications" Book Chapter, 16, 407–432, Nanoporous Solids, Recent Advances and Prospects, Elsevier 2008.

Chapter 7

Metal-Organic Framework Membranes for Hydrocarbon Separations

7.1. Introduction

Ethylene and propylene are one of the important industrial feed stocks in the petrochemical industry. The separation of these mixtures using distillation is highly energy intensive.[1,2] Low-energy separation technologies could provide very huge reductions in energy consumption. It has been estimated that ~1012 Btu/y could be reduced in energy consumption employing these low energy separation technologies.[3–5] Alternative less energy-intensive separation technologies, include membranes, adsorption, absorption and extraction. However, several limitations, have prevented the adoption of these technologies to replace or complement cryogenic distillation processes.[2] The main limitations of membrane separation technology for olefin/paraffin separation are their lack of selectivity, their narrow range of useful operation conditions and their typical high costs.[6–8] In principle, the separation of olefins and paraffins using polymeric membranes is quite challenging and difficult because of the similar molecular sizes, solubilities, physical and chemical properties. Polymeric membranes cannot achieve the permeability and separation selectivity combinations required for separating C_2H_4 from C_2H_6.[9–12] Futhermore, for most polymeric materials, the membrane stability under actual aggressive hydrocarbon feeds requires considerable attention.[13] Because of these concerns, olefin/paraffin separation applications might be one for which the benefits of novel membrane compositions may justify their higher costs.[13,14]

Metal-organic frameworks are considered as the potential candidates because of the highly accessible porosity which infers high fluxes, while the wide range of pore sizes and stability at harsh conditions would make it possible to tackle these difficult separations. A few types of MOFs have shown promising molecular sieve performances for hydrocarbon separations and simultaneously increased both propylene permeability and selectivity.

7.2. MOF membranes for olefin/paraffin separation

ZIF-8 membranes were first synthesized by Bux *et al.* on α-Al_2O_3 supports by the seeded-secondary solvothermal growth method and the resultant membranes showed promising separation performances with H_2/CH_4, H_2/C_2H_6 and H_2/C_3H_8 mixture selectivity of 15, 15 and 400 respectively.[15] Later, Pan *et al.* also synthesized a thin high quality ZIF-8 membranes on alumina supports at low temperature in aqueous solutions. These thin membranes showed excellent separation performance with a selectivity of about 10–167 for mixtures of ethane/propane, ethylene/propylene, ethylene/propane and propylene/propane.[16,17]

Lin's group[17] identified the separation mechanisms for propylene/propane over ZIF-8 membranes. Specifically this group found that propylene displayed higher diffusivity with smaller activation energy for diffusion than propane, consistent with the sizes of these two molecules with respect to the pore size of ZIF-8 membrane, leading to propylene selective membranes. Figure 7.1 shows SEM cross-section images of these ZIF-8 membranes displaying similar thickness of about 2.5 μm.[17] Top view images show that ZIF-8 crystals are more faceted with increasing synthesis temperature, suggesting enhanced crystallinity of the ZIF-8 layers as synthesis temperature increases.

Lai and coworkers also studied the propylene/propane separation performance and showed very high permeability of propylene up to 200 barrers and propylene/propane separation factor up to 50 at optimal separation conditions, well surpassing the "upper-bound trade-off".[18]

Kwon *et al.* developed a one-step *in situ* synthesis technique for high-quality MOF membranes based on the concept of counter diffusion which enabled the rapid preparation of well-intergrown ZIF-8 membranes with excellent microstructure.[19] These ZIF-8 membranes showed an excellent separation performance of a propylene/propane (50/50) mixture with a selectivity of ~55 and propylene permeability of 100 Barrer. Figure 7.2

Figure 7.1. SEM images of top and cross-section-views of ZIF-8 membranes prepared at 30°C (A and B), 80°C (C and D), and 120°C (E and F). (Reprinted from Ref. 17 with permission from Elsevier.)

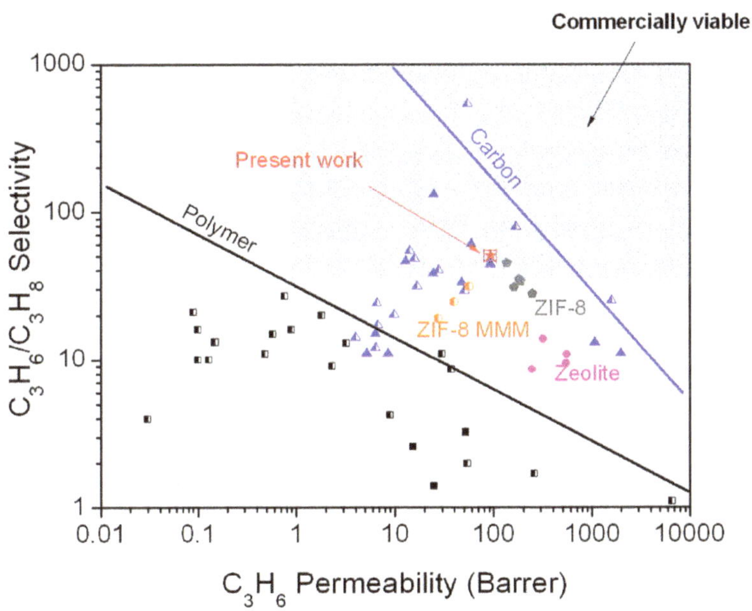

Figure 7.2. Comparison of propylene/propane separation performance of ZIF-8 membranes vs other reported membranes in the literature. (Reprinted from Ref. 19 with permission from American Chemical Society.)

compares the separation performance of ZIF-8 membranes vs other membranes reported in literature for propylene/propane separation.[19] It is clear that ZIF-8 membranes notably outperform both polymeric and zeolite membranes with respect to the separation factor and the propylene permeability. Furthermore, ZIF-8 membranes are close to the upper bound of carbon membranes while meeting the proposed requirement of minimum permeability of 1 Barrer and selectivity of 35 for potential commercial application.[19]

Brown *et al.* prepared ZIF-90 membranes on the outer surface of the hollow fiber supports and exhibit good separation properties for linear over cyclic hydrocarbons, as well as gas permeation selectivities higher than Knudsen values.[20] These membranes were selective for the linear n-hexane over cyclic hydrocarbons such as benzene and cyclohexane. Nair and coworkers prepared membranes using highly scalable interfacial microfluidic membrane processing for the first time and showed high gas separation performance with H_2/C_3H_8 and C_3H_6/C_3H_8 separation factors as high as 370 and 12, respectively.[21] Overall, MOF-based membranes showed very high hydrocarbon separation performances as compared to polymer and zeolite membranes with respect to both the separation factor and permeability. ZIF-8 was used extensively compared to the other MOFs for separating the hydrocarbons and successfully produced highly effective membranes.

7.3. ZIF-8 membranes for CO_2, hydrocarbons, and noble gases separations

ZIF-8 membranes have been widely employed (likely the most employed MOF membrane composition) for several relevant gas mixtures. Their unique pore size lying between the sizes of several industrially relevant gas molecules including have made this MOF composition highly appealing to separate several gas mixture pairs. In addition, since 2009 (first reports on continuous ZIF-8 membranes) considerable amount of work has been done to improve the quality (integrity and reproducibility) of these membranes. Table 7.1 summarizes *representative* examples of ZIF-8 membranes that have been successfully employed to separate molecular gas mixtures.

Table 7.1. Representative examples of ZIF-8 membranes employed for the separation of industrially relevant gas mixtures. References are listed chronologically.

Reference	Separation Selectivity α	Permeance (mol/m^2sPa)	General Separation Conditions
[22] Bux *et al.* 2009	$H_2/O_2 = 6^a$	H_2: $6.0 \times 10^{-8\,a}$, 5.1×10^{-8}	25°C $P_{per} = 0.1$ MPa equimolar H_2:CH_4 mixture
	$H_2/N_2 = 11.6^a$	CO_2: 1.3×10^{-8}	
	$CO_2/CH_4 = 2.8^a$	O_2: 1.0×10^{-8}	
	$N_2/CH_4 = 1.1^a$	N_2: 5.2×10^{-9}	
	$H_2/CH_4 = 13.0^a$, 11.2^b	CH_4: 4.6×10^{-9}	
[23] Venna & Carreon, 2010	$CO_2/CH_4 = 6.5–10.0^b$	CO_2: $1.69–2.73 \times 10^{-5}$	25°C $\Delta P = 0.1$ MPa equimolar CO_2:CH_4 mixture
		CH_4: $2.42–5.17 \times 10^{-6}$	
[24] Bux *et al.* 2010	$CO_2/CH_4 = 3.5^{a\&b}$	CO_2: 2.7×10^{-8}	22°C $P_{feed} = 0.14$ MPa $P_{per} = 0.10$ MPa equimolar CO_2:CH_4 mixture
		CH_4: 7.7×10^{-9}	
[25] McCarthy *et al.* 2010	$H_2/O_2 = 3.3^a$	H_2: 1.73×10^{-7}	25°C $P_{feed} = 0.1$ MPa
	$H_2/N_2 = 12^a$	CO_2: 4.45×10^{-8}	
	$CO_2/CH_4 = 3.0^a$	O_2: 5.22×10^{-8}	
	$N_2/CH_4 = 1.2^a$	N_2: 1.5×10^{-8}	
	$H_2/CH_4 = 13^a$	CH_4: 1.3×10^{-8}	

(*Continued*)

Table 7.1. (*Continued*)

Reference	Separation Selectivity α	Permeance (mol/m^2sPa)	General Separation Conditions
[15] Bux *et al.* 2011a	$H_2/CO_2 = 6^b$	H_2: 1.0×10^{-7}	25°C $P_{per} = 0.1$ MPa FR = 80–100 ml/min equimolar gas mixture
		CO_2: 1.6×10^{-8}	
	$H_2/CH_4 = 15^b$	H_2: 1.0×10^{-7}	
		CH_4: 6.7×10^{-8}	
	$H_2/C_2H_6 = 15^b$	H_2: 6.0×10^{-8}	
		C_2H_6: 4.0×10^{-9}	
	$H_2/C_3H_8 = 372^b$	H_2: 6.7×10^{-8}	
		C_3H_8: 1.8×10^{-10}	
[26] Bux *et al.* 2011b	$C_2H_4/C_2H_6 = 4.2^a, 2.8^b$	C_2H_4: 1.7×10^{-8}	25°C $P^a_{feed} = 0.6$ MPa $P^b_{feed} = 0.1$ MPa $P_{per} = 0.1$ MPa equimolar gas mixture
		C_2H_4: 1.7×10^{-8}	
		C_2H_6: 0.4×10^{-8}	
		C_2H_6: 0.6×10^{-8}	
[27] Hertäg *et al.* 2011	$H_2/CH_4 = 20.28^a$	H_2: 1.54×10^{-7}	25°C $P_{feed} = 0.1$MPa equimolar gas mixture
		CH_4: 7.6×10^{-9}	
	$H_2/CH_4 = 15.99^b$	H_2: 1.33×10^{-7}	
		CH_4: 8.3×10^{-9}	

Table 7.1. (*Continued*)

Reference	Separation Selectivity α	Permeance (mol/m^{2}sPa)	General Separation Conditions
[16] Pan and Lai 2011	$H_2/C_3H_8 = 545^b$	H_2: 4.36×10^{-7}	23°C equimolar gas mixture
		C_3H_8: 8×10^{-10}	
	$H_2/C_3H_6 = 23^b$	H_2: 3.84×10^{-7}	
		C_3H_6: 1.64×10^{-8}	
	$C_2H_6/C_3H_8 = 80^b$	C_2H_6: 7.23×10^{-8}	
		C_3H_8: 9×10^{-10}	
	$C_2H_4/C_3H_6 = 10^b$	C_2H_4: 1.47×10^{-7}	
		C_3H_6: 1.45×10^{-8}	
	$C_2H_4/C_3H_8 = 167^b$	C_2H_4: 1.50×10^{-7}	
		C_3H_8: 9×10^{-10}	
[28] Xu and *et al.* 2011	$H_2/CO_2 = 7.1^b$	H_2: 1.7×10^{-7}	25°C $P_{per} = 0.1$ MPa 45% H_2
		CO_2: 2.4×10^{-8}	
[29] Yao *et al.* 2011	$H_2/N_2 = 1.6–4.3^a$	H_2: $3.80–22.01 \times 10^{-6}$	
		N_2: $0.46–13.56 \times 10^{-8}$	
[18] Pan *et al.* 2012	$C_3H_6/C_3H_8 = 45^b$	C_3H_6: 2.06×10^{-8}	22°C equimolar gas mixture
		C_3H_8: 4.6×10^{-10}	
[19] Kwon *et al.* 2013	$C_3H_6/C_3H_8 = 50^b$	C_3H_6: 2.1×10^{-8}	room temp equimolar gas mixture
		C_3H_8: 4.2×10^{-10}	

(*Continued*)

Table 7.1. *(Continued)*

Reference	Separation Selectivity α	Permeance (mol/m^{2}sPa)	General Separation Conditions
[21] Brown *et al.* 2014	$H_2/C_3H_8 = 409$[b]	H_2: 2.9×10^{-7}	120°C equimolar gas mixture
		C_3H_8: 7×10^{-10}	
	$C_3H_6/C_3H_8 = 14.5$[b]	C_3H_6: 8.3×10^{-9}	25°C equimolar gas mixture
		C_3H_8: 5.7×10^{-10}	
[30] Cacho *et al.* 2014	$H_2/N_2 = 12.4$[b]	H_2: 2.02×10^{-7}	35°C $\Delta P = 0$ FR = 180 ml/min
		N_2: 1.7×10^{-8}	
	$H_2/CH_4 = 10.5$[b]	H_2: 1.9×10^{-7}	
		CH_4: 1.8×10^{-8}	
	$H_2/CH_4 = 11.7$[b]	H_2: 2.7×10^{-7}	80°C $\Delta P = 0$ FR = 180 ml/min
		CH_4: 2.3×10^{-8}	
[31] Hara *et al.* 2014	$C_3H_6/C_3H_8 = 59$[a]	C_3H_6: 2.4×10^{-9}	25°C $\Delta P = 0.1$ MPa
		C_3H_6: 2.4×10^{-9}	

Table 7.1. (*Continued*)

Reference	Separation Selectivity α	Permeance (mol/m^{2}sPa)	General Separation Conditions
[17] Liu *et al.* 2014	$C_3H_6/C_3H_8 = 35^b$	C_3H_6: 1.0×10^{-8}	35°C $P_{feed} = 0.2$ MPa FR = 40 ml/min 15% C_3H_6
		C_3H_8: 2.9×10^{-10}	
	$C_3H_6/C_3H_8 = 28.8^a$	C_3H_6: 1.0×10^{-8}	35°C $P_{feed} = 0.1$ MPa $P_{per} = 0$ MPa FR= 40 ml/min
		C_3H_8: 2.8×10^{-10}	
	$H_2/He = 2.5^a$	He: 1.9×10^{-7}	35°C $P_{per} = 0.1$ MPa FR = 40 ml/min
	$H_2/CO_2 = 3.9^a$	H_2: 4.7×10^{-7}	
	$H_2/N_2 = 10.4^a$	CO_2: 1.2×10^{-7}	
	$CO_2/CH_4 = 2.9^a$	N_2: 4.5×10^{-8}	
	$N_2/CH_4 = 1.1^a$	CH_4: 4.2×10^{-8}	
	$H_2/SF_6 = 1620^a$	SF_6: 2.9×10^{-10}	
[32] Shekhah, *et al.* 2014	$H_2/He = 2.13^a$	He: 8.9×10^{-9}	35°C P = 0.2 MPa
	$H_2/N_2 = 10^a$	H_2: 1.9×10^{-8}	
	$H_2/O_2 = 3.88^a$	N_2: 1.9×10^{-9}	
	$H_2/CH_4 = 9.5^a$	O_2: 4.9×10^{-9}	
	$H_2/CO_2 = 4.6^a$	CH_4: 2.0×10^{-9}	
	$CO_2/CH_4 = 2.0^a$	CO_2: 4.1×10^{-9}	
	$H_2/C_2H_4 = 4.6^a$	C_2H_4: 4.1×10^{-9}	
	$C_2H_6/C_3H_6 = 0.3^a$	C_2H_6: 1.9×10^{-9}	
	$C_3H_6/C_3H_8 = 3.5^a$	C_3H_6: 6.0×10^{-10}	
	$C_3H_8/C_4H_{10} = 0.7^a$	C_3H_8: 1.7×10^{-10}	
		C_4H_{10}: 2.3×10^{-10}	

(*Continued*)

Table 7.1. (*Continued*)

Reference	Separation Selectivity α	Permeance (mol/m^2sPa)	General Separation Conditions
[33] Venna *et al.* 2014	$N_2/CH_4 = 0.96^b$	N_2: 7.8×10^{-7}	24°C $P_{per} = 0.1$ MPa $\Delta P = 0.1$ MPa FR = 100 ml/min
		CH_4: 8.2×10^{-7}	
	$CO_2/N_2 = 0.77^b$	CO_2: 3.7×10^{-7}	90°C $P_{per} = 0.1$ MPa $\Delta P = 0.1$ MPa FR = 100 ml/min
		N_2: 4.9×10^{-7}	
[34] Zhang *et al.* 2014	$H_2/CO_2 = 5.2^a, 4.6^b$	H_2: $2.1 \times 10^{-7\,a}$, 1.6×10^{-7}	$P_{per} = 0.1$ MPa $\Delta P = 0.6$–1.2 MPa FR = 100 ml/min
		CO_2: $3.9 \times 10^{-8\,a}$, 3.4×10^{-8}	
	$H_2/N_2 = 10.3^a, 8.2^b$	H_2: $2.05 \times 10^{-7\,a}$, 1.69×10^{-7}	$P_{per} = 0.1$ MPa $P_{feed} = 0.1$ MPa $\Delta P = 0$ FR = 100 ml/min and 50 ml/min
		N_2: $2.0 \times 10^{-8\,a}$, 2.1×10^{-8}	
	$H_2/CH_4 = 11.4^a, 9.8^b$	H_2: $2.05 \times 10^{-7\,a}$, 1.67×10^{-7}	
		CH_4: $2.8 \times 10^{-8\,a}$, 1.7×10^{-8}	

Table 7.1. (*Continued*)

Reference	Separation Selectivity α	Permeance (mol/m^2sPa)	General Separation Conditions
[35] Cacho *et al.* 2015	$H_2/CH_4 = 17.2^b$	H_2: $4.8 \pm 1.3 \times 10^{-9}$	35°C $\Delta P = 0$ FR = 10–30 ml/min equimolar gas mixture
		CH_4: $3 \pm 2 \times 10^{-10}$	
	$H_2/N_2 = 18.3^b$	H_2: $4.8 \pm 1.4 \times 10^{-9}$	
		N_2: $3 \pm 2 \times 10^{-10}$	
	$H_2/CO_2 = 2.6^b$	H_2: $4.6 \pm 1.4 \times 10^{-9}$	
		CO_2: $1.8 \pm 0.2 \times 10^{-9}$	
	$CO_2/CH_4 = 6.1^b$	CO_2: $1.5 \pm 3 \times 10^{-9}$	
		CH_4: $3 \pm 2 \times 10^{-10}$	
	$CO_2/N_2 = 7.1^b$	CO_2: $1.6 \pm 0.2 \times 10^{-9}$	
		N_2: $3 \pm 2 \times 10^{-10}$	
[36] Drobek *et al.* 2015	$H_2/CH_4 = 12.5^b$	H_2: 5.4×10^{-8}	100°C $\Delta P = 0.2$ MPa equimolar gas mixture
		CH_4: 4.3×10^{-9}	
	$H_2/CO_2 = 7.8^b$	H_2: 5.8×10^{-8}	
		CO_2: 7.4×10^{-9}	
	$H_e/CO_2 = 2.2^a$	He: 6.1×10^{-8}	25°C $\Delta P = 0.2$ MPa
	$CO_2/N_2 = 2.3^a$	CO_2: 2.8×10^{-8}	
	$CO_2/CH_4 = 2.8^a$	N_2: 1.2×10^{-8}	
	$N_2/CH_4 = 1.2^a$	CH_4: 1.0×10^{-8}	
	$H_e/SF_6 = 61^a$	SF_6: 8.9×10^{-10}	
[37] Isaeva *et al.* 2015	$He/CO_2 = 9.2^a$ $He/N_2 = 2.8^a$	He: 6.0×10^{-9}	room temp $P_{per} = \sim 0.001$ MPa $\Delta P = 0.05$–0.35 MPa
		N_2: 6.9×10^{-10}	
		CO_2: 6.5×10^{-10}	

(*Continued*)

Table 7.1. *(Continued)*

Reference	Separation Selectivity α	Permeance (mol/m^2sPa)	General Separation Conditions
[38] Kong *et al.* 2015	$H_2/CO_2 = 5.65$[a]	H_2: $6.70–7.22 \times 10^{-8}$	$\Delta P = 0.4–1.2$ MPa
	$H_2/O_2 = 6.09$[a]	CO_2: $1.18–1.63 \times 10^{-8}$	
	$H_2/N_2 = 7.15$[a]	O_2: $1.18–1.27 \times 10^{-8}$	
	$H_2/CH_4 = 7.89$[a]	N_2: $0.94–1.02 \times 10^{-8}$	
		CH_4: $0.87–0.92 \times 10^{-8}$	
[39] Kwon *et al.* 2015	$C_3H_6/C_3H_8 = 70.6$[b]	C_3H_6: 2.69×10^{-8}	FR = 100 ml/min equimolar gas mixture
		C_3H_8: 3.80×10^{-10}	
[40] Melgar *et al.* 2015	$H_2/CO_2 = 20.29$[b]	H_2: 4.23×10^{-7}	equimolar gas mixture
		CO_2: 2.08×10^{-8}	
[41] Pan *et al.* 2015	$C_3H_6/C_3H_8 = 89$[b]	C_3H_6: 7.8×10^{-9}	22–50°C FR= 100 ml/min equimolar gas mixture
		C_3H_8: 8.8×10^{-11}	
[42] Cacho *et al.* 2016	$H_2/CO_2 = 4.1$[b]	H_2: 1.2×10^{-8}	35°C $\Delta P = 0$ FR= 10 ml/min equimolar gas mixture
		CO_2: 2.9×10^{-9}	
	$H_2/CO_2 = 7.4$[b]	H_2: 6.6×10^{-8}	150°C $\Delta P = 0$ FR = 10 ml/min equimolar gas mixture
		CO_2: 8.9×10^{-9}	

Table 7.1. (*Continued*)

Reference	Separation Selectivity α	Permeance (mol/m^2sPa)	General Separation Conditions
[43] Shamsaei *et al.* 2016	$C_3H_6/C_3H_8 = 27.8^a$	C_3H_6: 7.5×10^{-9}	room temp $\Delta P = 0.1$ MPa
		C_3H_8: 2.7×10^{-10}	
	$H_2/ C_3H_8 = 2259^a$	H_2: 6.1×10^{-7}	
		C_3H_8: 2.7×10^{-10}	
[44] Lai *et al.* 2016	$CO_2/CH_4 = 2.26^a$	CO_2: 1.31×10^{-7}	30°C $\Delta P = 0.1$ MPa $FR_f = 200$ ml/min
		CH_4: 5.8×10^{-8}	
	$CO_2/CH_4 = 5.07^b$	CO_2: 6.6×10^{-8}	30°C $\Delta P = 0.1$ MPa FR = 200 ml/min 15% CO_2
		CH_4: 1.3×10^{-8}	
[45] Yu *et al.* 2016	$C_3H_6/C_3H_8 = 61^b$	C_3H_6: 6.2×10^{-9}	FR = 50 ml/min equimolar gas mixture
		C_3H_8: 1.02×10^{-10}	

(*Continued*)

Table 7.1. *(Continued)*

Reference	Separation Selectivity α	Permeance (mol/m^2sPa)	General Separation Conditions
[46] Wu *et al.* 2017	$Kr/Xe = 5.9–16.1^b$	Kr: $0.33–1.7 \times 10^{-8}$ Xe: $0.05–0.14 \times 10^{-8}$	FR = 70 ml/min Feed composition (mol) 92:8 Kr/Xe
[47] Zhang *et al.* 2017	$He/SF_6 = 184^a$ (representative example). Others can be extracted from the permeance column	H_2: $4 \times 10^{-7\,a}$; He: $1.5 \times 10^{-7\,a}$; N_2: $0.4 \times 10^{-7\,a}$ O_2: $0.8 \times 10^{-7\,a}$; CO_2: $0.85 \times 10^{-7\,a}$; SF_6: $0.34 \times 10^{-9\,a}$	N/A
[48] James *et al.* 2017	$H_2/CO_2 \sim 5.2^b$	H_2: $\sim 2 \times 10^{-6\,b}$ CO_2: $\sim 0.38 \times 10^{-6\,b}$	FR = 25 ml/min. Pressure 1atm. T-25–300°C. Equimolar feed composition
[49] James *et al.* 2017	$C_2H_4/C_2H_6 \sim 3^a$ $C_2H_4/C_2H_6 \sim 2^b$	C_2H_4: $100 \times 10^{-9\,a}$ C_2H_4: $90 \times 10^{-9\,b}$ C_2H_6: $40 \times 10^{-9\,a}$ C_2H_6: $50 \times 10^{-9\,b}$	FR = 100 ml/min. Pressure 1–4 atm. T-25–100°C. Equimolar feed composition

FR = flow rate, [a]indicates ideal selectivity; [b]indicates gas mixture selectivity

References

1. Gottschlich, R. Energy Minimization of Separation Processes using Conventional/Membrane Hybrid Systems. Washington DC: US Dept. of Energy; 1990.
2. Robinson, S., Jubin, R. & Choate, B. Materials for Separation Technology: Energy and Emmision Reduction Opportunities. Washington DC: US Dept. of Energy; 2005.
3. Eldridge, R. B. Olcfin paraffin separation technology: A review. *Ind. Eng. Chem. Res.* **32**, 2208–2212 (1993).
4. Caballero, J. A., Grossmann, I. E., Keyvani M. & Lenz, E. S. Design of hybrid distillation-vapor membrane separation systems. *Ind. Eng. Chem. Res.* **48**, 9151–9162 (2009).
5. Rungta, M., Xu, L. & Koros, W. J. Carbon molecular sieve dense film membranes derived from MatrimidV R for ethylene/ethane separation. *Carbon.* **50**, 1488–1502 (2012).
6. Baker, R. W. Future directions of membrane gas separation technology. *Ind. Eng. Chem. Res.* **41**, 1393–1411 (2002).
7. Ozokwelu, D. Hybrid Separations/Distillation Technology: Research Opportunities for Energy and Emissions Reduction. Washington DC: US Dept. of Energy; 2005.
8. Koros, W. J. Evolving beyond the thermal age of separation processes: Membranes can lead the way. *AIChE J.* **50**, 2326–2334 (2004).
9. Chan, S. S., Wang, R., Chung, T. S. & Liu, Y. C-2 and C-3 hydrocarbon separations in poly(1,5-naphthalene-2,20-bis(3,4-phthalic) hexauoropropane) diimide (6FDA-1,5-NDA) dense membranes. *J. Membr. Sci.* **210**, 55–64 (2002).
10. Chan, S. S., Chung, T.-S., Liu, Y. & Wang, R. Gas and hydrocarbon (C2 and C3) transport properties of co-polyimides synthesized from 6FDA and 1,5-NDA (naphthalene)/durene diamines. *J. Membr. Sci.* **218**, 235–245 (2003).
11. Tanaka, K., Taguchi, A., Hao, J., Kita, H. & Okamoto, K. Permeation and separation properties of polyimide membranes to olefins and paraffins. *J. Membr. Sci.* **121**, 197–207 (1996).
12. Staudt-Bickel, C. & Koros, W. J. Olefin/paraffin gas separations with 6FDA-based polyimide membranes. *J. Membr. Sci.* **170**, 205–214 (2000).
13. Rungta, M., Zhang, C., Koros, W. J. & Xu, L. Membrane-based ethylene/ethane separation: The upper bound and beyond. *AIChE Journal* **59**(9), 3475–3479 (2013).
14. Koros, W. J. & Mahajan, R. Pushing the limits on possibilities for large scale gas separation: Which strategies? *J. Membr. Sci.* **175**, 181–196 (2000).
15. Bux, H., Feldhoff, A., Cravillon, J., Wiebcke, M., Li, Y. S. & Caro, J. Oriented zeolitic imidazolate framework-8 membrane with sharp H_2/C_3H_8 molecular sieve separation. *Chem. Mater.* **23**, 2262–2269 (2011).
16. Pan, Y. & Lai, Z. Sharp separation of C_2/C_3 hydrocarbon mixtures by zeolitic imidazolate framework-8 (ZIF-8) membranes synthesized in aqueous solutions. *Chem. Commun. (Camb).* **47**, 10275–10277 (2011).
17. Liu, D., Ma, X., Xi, H. & Lin, Y. S. Gas transport properties and propylene/propane separation characteristics of ZIF-8 membranes. *J. Memb. Sci.* **451**, 85–93 (2014).
18. Pan, Y., Li, T., Lestari, G. & Lai, Z. Effective separation of propylene/propane binary mixtures by ZIF-8 membranes. *J. Memb. Sci.* **390–391**, 93–98 (2012).

19. Kwon, H. T. & Jeong, H.-K. *In situ* synthesis of thin zeolitic-imidazolate framework ZIF-8 membranes exhibiting exceptionally high propylene/propane separation. *J. Am. Chem. Soc.* **135**, 10763–10768 (2013).
20. Brown, A. J., Johnson, J. R., Lydon, M. E., Koros, W. J., Jones, C. W. & Nair, S. Continuous polycrystalline zeolitic imidazolate framework-90 membranes on polymeric hollow fibers. *Angew. Chem. Int. Ed. Engl.* **51**, 10615–10618 (2012).
21. Brown, A. J. Brunelli, N. A., Eum, K., Rashidi, F., Johnson, J. R., Koros, W. J., Jones, C. W. & Nair, S. Separation membranes. Interfacial microfluidic processing of metal-organic framework hollow fiber membranes. *Science* **345**, 72–75 (2014).
22. Bux, H., Liang, F., Li, Y., Cravillon, J., Wiebcke, M. & Caro, J. Zeolitic imidazolate framework membrane with molecular sieving properties by microwave-assisted solvothermal synthesis. *Journal of the American Chemical Society* **131**(44), 16000–16001 (2009).
23. Venna, S. R. & Carreon, M. A. Highly permeable zeolite imidazolate framework-8 membranes for CO_2/CH_4 separation. *Journal of the American Chemical Society* **132**(1), 76–78 (2010).
24. Bux, H., Chmelik, C., van Baten, J. M., Krishna, R. & Caro, J. Novel MOF-membrane for molecular sieving predicted by IR-diffusion studies and molecular modeling. *Advanced Materials* **22**(42), 4741–4743 (2010).
25. McCarthy, M. C., Varela-Guerrero, V., MPanett, G. V. & Jeong, H. K. Synthesis of zeolitic imidazolate framework films and membranes with controlled microstructures. *Langmuir* **26**(18), 14636–14641 (2010).
26. Bux, H., Chmelik, C., Krishna, R. & Caro, J. Ethene/ethane separation by the MOF membrane ZIF-8: molecular correlation of permeation, adsorption, diffusion. *Journal of Membrane Science* **369**(1), 284–289 (2011b).
27. Hertäg, L., Bux, H., Caro, J., Chmelik, C., Remsungnen, T., Knauth, M. & Fritzsche, S. Diffusion of CH_4 and H_2 in ZIF-8. *Journal of Membrane Science* **377**(1), 36–41 (2011).
28. Xu, G., Yao, J., Wang, K., He, L., Webley, P. A., Chen, C. S. & Wang, H. Preparation of ZIF-8 membranes supported on ceramic hollow fibers from a concentrated synthesis gel. *Journal of Membrane Science* **385**, 187–193 (2011).
29. Yao, J., Dong, D., Li, D., He, L., Xu, G. & Wang, H. Contra-diffusion synthesis of ZIF-8 films on a polymer substrate. *Chemical Communications* **47**(9), 2559–2561 (2011).
30. Cacho-Bailo, F., Seoane, B., Téllez, C. & Coronas, J. ZIF-8 continuous membrane on porous polysulfone for hydrogen separation. *Journal of Membrane Science* **464**, 119–126 (2014).
31. Hara, N., Yoshimune, M., Negishi, H., Haraya, K., Hara, S. & Yamaguchi, T. Diffusive separation of propylene/propane with ZIF-8 membranes. *Journal of Membrane Science* **450**, 215–223 (2014).
32. Shekhah, O., Swaidan, R., Belmabkhout, Y., Du Plessis, M., Jacobs, T., Barbour, L. J., Pinnau, I. & Eddaoudi, M. (2014). The liquid phase epitaxy approach for the successful construction of ultra-thin and defect-free ZIF-8 membranes: pure and mixed gas transport study. *Chemical Communications* **50**(17), 2089–2092 (2014).

33. Venna, S. R., Zhu, M., Li, S. & Carreon, M. A. Knudsen diffusion through ZIF-8 membranes synthesized by secondary seeded growth. *Journal of Porous Materials* **21**(2), 235–240 (2014).
34. Zhang, X., Liu, Y., Li, S., Kong, L., Liu, H., Li, Y., Han, W., Yeung, K. L., Zhu, W., Qiu, J. & Qiu, J. New membrane architecture with high performance: ZIF-8 membrane supported on vertically aligned ZnO nanorods for gas permeation and separation. *Chemistry of Materials* **26**(5), 1975–1981 (2014).
35. Cacho-Bailo, F., Catalan-Aguirre, S., Etxeberria-Benavides, M., Karvan, O., Sebastian, V., Tellez, C. & Coronas, J. Metal-organic framework membranes on the inner-side of a polymeric hollow fiber by microfluidic synthesis. *Journal of Membrane Science* **476**, 277–285 (2015).
36. Drobek, M., Bechelany, M., Vallicari, C., Chaaya, A. A., Charmette, C., Salvador-Levehang, C. & Julbe, A. An innovative approach for the preparation of confined ZIF-8 membranes by conversion of ZnO ALD layers. *Journal of Membrane Science* **475**, 39–46 (2015).
37. Isaeva, V. I., MPakova, M. I., Kustov, L. M., Syrtsova, D. A., Efimova, E. A. & Teplyakov, V. V. *In situ* synthesis of novel ZIF-8 membranes on polymeric and inorganic supports. *Journal of Materials Chemistry* **A3**(14), 7469–7476 (2015).
38. Kong, L., Zhang, X., Liu, H., Wang, T. & Qiu, J. Preparation of ZIF-8 membranes supported on macroporous carbon tubes via a dipcoating–rubbing method. *Journal of Physics and Chemistry of Solids* **77**, 23–29 (2015).
39. Kwon, H. T. & Jeong, H. K. Improving propylene/propane separation performance of Zeolitic-Imidazolate framework ZIF-8 Membranes. *Chemical Engineering Science* **124**, 20–26 (2015).
40. Melgar, V. M. A., Ahn, H., Kim, J. & Othman, M. R. Highly selective micro-porous ZIF-8 membranes prepared by rapid electrospray deposition. *Journal of Industrial and Engineering Chemistry* **21**, 575–579 (2015).
41. Pan, Y., Liu, W., Zhao, Y., Wang, C. & Lai, Z. Improved ZIF-8 membrane: Effect of activation procedure and determination of diffusivities of light hydrocarbons. *Journal of Membrane Science* **493**, 88–96 (2015).
42. Cacho-Bailo, F., Matito-Martos, I., Perez-Carbajo, J., Etxeberría-Benavides, M., Karvan, O., Sebastián, V., Calero, S., Téllez, C. & Coronas, J. On the molecular mechanisms for the H_2/CO_2 separation performance of zeolite imidazolate framework two-layered membranes. *Chemical Science* (2017).
43. Lai, L. S., Yeong, Y. F., Chew, T. L., Lau, K. K. & Azmi, M. S. (2016). CO_2 and CH_4 gas permeation study via zeolitic imidazolate framework (ZIF)-8 membrane. *Journal of Natural Gas Science and Engineering* **34**, 509–519 (2016).
44. Shamsaei, E., Lin, X., Low, Z. X., Abbasi, Z., Hu, Y., Liu, J. Z. & Wang, H. Aqueous phase synthesis of ZIF-8 membrane with controllable location on an asymmetrically porous polymer substrate. *ACS Applied Materials & Interfaces* **8**(9), 6236–6244 (2016).
45. Yu, J., Pan, Y., Wang, C. & Lai, Z. ZIF-8 membranes with improved reproducibility fabricated from sputter-coated ZnO/alumina supports. *Chemical Engineering Science* **141**, 119–124 (2016).

46. Wu, T., Feng, X., Elsaidi, S. K., Thallapally, P. K. & Carreon, M. A. Zeolitic Imidazolate Framework-8 (ZIF-8) membranes for Kr/Xe separation, *Industrial & Engineering Chemistry Research* **56**, 1682–1686 (2017).
47. Zhang, H., James, J., Zhao, M., Yao, Y., Zhang, Y., Zhang, B. & Lin, Y.S. Improving hydrostability of ZIF-8 membranes via surface ligand exchange. *Journal of Membrane Science* **532**, 1–8 (2017).
48. James, J. B. & Lin, Y. S. Thermal stability of ZIF-8 membranes for gas separations. *Journal of Membrane Science* **532**, 9–19 (2017).
49. James, J. B, Wang, J., Meng, L. & Lin, Y. S. ZIF-8 Membrane ethylene/ethane transport characteristics in single and binary gas mixtures. *Industrial & Engineering Chemistry Research* **56**(26), 7567–7575 (2017).

Chapter 8

Metal-Organic Framework Membranes for Xenon Capture

8.1. Kr/Xe separation

8.1.1. *Introduction*

The separation of Krypton (Kr) from Xenon (Xe) is a relevant problem for nuclear industries. Separating Kr from Xe is a critical step in removing radioactive Kr during treatment of spent nuclear fuel. The conventional method to separate these gases is fractional distillation at cryogenic temperatures, which is an energy intensive process. In addition, even after cryogenic distillation, trace levels of radioactive Kr in the Xe-rich phase are too high to permit for further use. In this respect, membrane technology could play a key role in making this separation less energy intensive and therefore economically feasible. Membrane separation processes have several advantages over conventional fractional distillation: membrane separation is energy-saving since it does not involve phase transformation; the required membrane process equipment is also simple, easy to operate, control and scale-up. Metal-organic frameworks (MOFs) have emerged as a suitable type of crystalline porous materials which combine highly desirable properties, such as uniform micropores, high surface areas, and exceptional thermal and chemical stability, making them ideal candidates for Kr/Xe separation.

8.1.2. *Nuclear energy generalities*

Nuclear power represents a powerful, and reliable alternative energy source that emits very little CO_2. A coal plant would emit more than one million tons of CO_2 to produce the same amount of electricity that a nuclear power plant could with 22 tons of uranium. The typical nuclear plant even emits fewer radioactive particles than a traditional coal-burning plant, suggesting additional environmental benefits. Nuclear power compares very favorably with other clean energy technologies.[1] For instance, unlike wind and solar, nuclear power can be produced continuously, regardless of the weather. Nuclear power provides about 70% of the United States' clean electricity. More than 11% of the world's electricity comes from continuous, reliable base-load nuclear power.

8.1.3. *Nuclear power waste storage and fuel reprocessing*

Nuclear power does have drawbacks, and one of the greatest is its radioactive waste, which amounts to thousands of tons per year. Uranium and plutonium, the most common nuclear fuels, have created 76,430 metric tons of radioactive nuclear waste in the US over the last 40 years.[2] The US government has sought to create a repository for the radioactive waste created by nuclear power generation. Starting in 1987, the Yucca Mountain site in Nevada was studied as a nuclear waste repository. Alone, however, geological repositories cannot solve the nuclear waste problem. Nuclear waste reprocessing is gaining support as a complement to nuclear waste repositories. Reprocessing reduces the amount of high-level waste, including plutonium and uranium, and enables more energy to be extracted from the fuel. Non-usable low-level waste must still be sent to a repository. Some European countries, along with Russia, China, and Japan, have adopted policies to reprocess spent nuclear fuel.[3] Partnering with the International Framework for Nuclear Energy Cooperation (IFNEC), the United States is developing a closed nuclear fuel cycle that recycles nuclear waste and extracts additional energy and economic value.[4] A main research focus in spent fuel reprocessing is capturing the volatile "radionuclides" in off-gases, which are more difficult to manage than solid or liquid radionuclides. The off-gases from fuel reprocessing contain NO, NO_2, ^{131}I, Xe, ^{85}Kr, $^{14}CO_2$, tritium (as ^3H_2O).[5] In the future, to meet likely US regulatory requirements, reprocessing facilities must capture volatile radionuclides. Currently, some commercially available technologies are applied in nuclear reprocessing plants in Europe, Russia and Japan to capture radionuclides (^{131}I and/or tritium), but the capture and separation of Kr and Xe is still a challenge in nuclear reprocessing plants.[6]

8.1.4. *The significance of Kr/Xe separation*

Krypton and Xenon must be captured and separated from spent fuel during reprocessing due to the relatively long half-life of ^{85}Kr (10.8 year), which accounts for 0.2% of the Kr isotopes and decays into rubidium-85. ^{85}Kr emits beta radiation (99.6%) and gamma radiation (0.4%),[7] direct exposure to which harms human health and increases in cancer risk.[81] Xenon, however, does not pose the same radiation risks. After a relatively short period of time, ^{135}Xe (half-life of 9.2 hours) decays into stable isotope(s). However, as noble gases, Xe and ^{85}Kr are difficult to separate and must be removed from spent fuel together.

The radioactive ^{85}Kr must be stored for 110 years before atmospheric release.[10] After Xe and ^{85}Kr have been captured, they should be separated. Effectively separating these gases would reduce storage costs since Xe is generally present at 10 times the concentration of Kr and will allow for the economic use of the gases, especially Xe, which has many applications.

Kr captured from off-gases can be used as a radionuclide owing to the presence of ^{85}Kr. Kr is used as self-luminescent light sources, and in leak detectors, thickness gauges, and static eliminators.[11–14] However, ^{85}Kr can only be used in limited capacities due to the biological effects of radiation. The value of Kr from spent fuel from 1,000 MW Nuclear plant for 1 year is \$19,164 (308.18 gr in total, 0.2 % isotope, \$616 per metric ton commercially).[15]

Xe has several important commercial applications. If the Xe off-gas can be decontaminated to a very high degree, it can be used for commercial purposes. Recovered Xe can be used as a non-radioactive material in scintillation devices,[16] long lived, high intensity light bulbs,[17,18] and special electronic equipment.[19] Xe can also be applied in medical capacities, especially as a surgical anaesthetic with a short induction period, allowing for a full recovery within minutes.[20–22] The value of Xe from spent fuel from a 1,000 MW Nuclear plant for 1 year is \$281,546 (5,322 gr, \$9,050 per metric ton commercially).[15]

8.1.5. *Conventional methods to separate Kr and Xe*

Kr and Xe need to be physically separated because they are inert. The physical properties of Kr and Xe are shown in Table 8.1. The current sources

Table 8.1. Physical properties of Krypton and Xenon. (Taken from Ref. 23).

Physical properties	Kr	Xe
Kinetic diameter (nm)	0.366	0.405
Polarizability (10^{-25} cm^3)	24.84	40.44
Dipole moment (esu cm)	0	0
Quadruple moment (esu cm^2)	0	0

of Xe and Kr are very limited, owing to the prevailing separation method. Currently, a ~ 12:1 mixture of Xe and Kr is captured from the liquefaction of air.[22] The benchmark technology to separate Kr and Xe is cryogenic distillation process. This method of producing Kr and Xe is extremely energy-intensive, making the products very expensive. Apart from the cryogenic method, physical adsorption and membrane separation methods are emerging. These Xe/Kr separation methods are described in detail below.

Cryogenic Distillation

Cryogenic distillation is the most mature noble gas separation technology.[24] The technique relies on the difference in boiling points between Xe and Kr (−108.12°C for Xe, −153.22°C for Kr).[6] The cryogenic method can be applied to separate Xe and Kr from air or from spent nuclear fuel off-gases. For nuclear reprocessing plants, off-gases need to be pre-treated to remove impurities such as C_2H_2, NO_X and O_2.[25] Commercially, ~99.9% pure Kr and Xe can be obtained using cryogenic separation of Kr and Xe from air.[24] This method's application in nuclear fuel reprocessing plants has been studied for several decades in the US[26] and Japan.[24] To meet anticipated regulatory ^{85}Kr emission requirements, the optimum Kr decontamination factor should reach 100–1000.[24]

Cryogenic distillation methods for both air distillation and nuclear reprocessing plants are highly developed. Abe *et al.* reported the first cryogenic distillation system that can reach a concentration of Kr at 10^{-12} mol/mol in Xe.[27] The Xe collection efficiency was 99%. However, the process speed of 0.6 kg Xe per hour limited the application on a ton-scale pilot. After several years of work, the process speed was improved from 0.6 kg/h[28] to 5 kg/h[29] and 6.5 kg/h.[30] The system designed in 2016 was capable of processing 3.3 tons of Xe with an extremely low Kr contamination of 10^{-15} mol/mol.[30]

The cryogenic distillation process requires an expensive capital outlay and intensive energy inputs. The selling price of high-purity Xe from cryogenic distillation was $5,000/kg in 2014.[31] A safety concern with cryogenic distillation is the formation of ozone from oxygen at cryogenic temperatures caused by ^{85}Kr radiation.[32] Due to high costs and ozone concerns, a safer, more cost-effective method for separating Xe/Kr from off-gas streams is necessary. The optimal process would separate the noble gases at ambient temperature and pressure.

Dissolution in Fluorocarbon

Another approach to separate Kr from Xe is dissolution in fluorocarbon, which separates these gases through solubility rather than boiling point differences. The solubility of Kr/Xe varies by temperature and pressure based on the solvent. A selective adsorption process using CCl_2F_2 (commercial name: Freon-12 or R-12) as a solvent can reach Kr separation factors of up to 1,000 and removal efficiencies of 99.9%.[33] This technology requires capital equipment and operating costs similar to the cryogenic process.[34] However, the operating temperature is about −30°C, leading to lower refrigeration costs. The high operating pressure at around 2.8 MPa increases potential solvent and product loss through process leakage, volatilization and radiolysis degradation.[24] Moreover, the usage of CCl_2F_2 was forbidden globally in 2010 owing to concerns about its atmospheric ozone depletion effects.[35] The technology has never been tested at a pilot scale.[24]

Physical Adsorption

Physical adsorption is an alternative, potentially energy- and cost-saving Xe/Kr separation process. A broad range of task-specific porous materials have been investigated for Kr/Xe separation, including activated carbon,[36–44] zeolites[45–54] metal-organic frameworks[55–72] and porous organic cages.[73–77]

Selective adsorption of Kr and Xe on activated charcoal was investigated since the beginning of the middle 20th century and has been developed in recent years. Activated charcoal is a cheap carbon material with a high surface area[36] that has been used to separate Kr/Xe mixtures. Activated charcoal activation process produces abundant active sites beneficial to adsorption due to increased free suspended carbon bonds.[37] Unfortunately, additional processes are necessary to separate Kr from activated carbon, making activated carbon's use as an off-gas adsorbent inefficient. Nevertheless, because of activated carbon's inexpensiveness and adsorbent capacity, it is often used as a benchmark to compare the performance of other adsorbents.[36] In 1999, Munakata *et al.* showed that activated charcoal has greater adsorption capacity for both Kr and Xe than the molecular sieve 5A (MS5A) adsorbents.[38] Munakata *et al.* showed that some carbon-based adsorbents, including Ambersorb 572, Ryujyo-Sirasagi G2x4/6-3 and Molsievon X2M4/6 have large adsorption capacities for Kr under PSA.[39] Compared to different zeolites and activated carbon adsorbents, the activated charcoal Ambersorb 572 was found to have the highest adsorbent capacity for Kr,[40] reported at 2.2×10^{-6} mol/g[41] at 0°C, and 30 Pa. AUK porous carbon was reported with Kr capacity of $3.2 \times 10 - 3$ mol/g at 2,981 Pa at −96°C, and separation decreased as temperature increased.[37] Unfortunately, activated carbon is susceptible to combustion with O_2 or NOx in the feed stream.[36]

Single-walled carbon nanotubes (SWCNTs) can act as adsorbents for Kr and Xe due to their high surface area, uniform pore size and thermal stability. Muris *et al.* investigated the capture of methane and Kr on the inner channels and exterior walls of carbon nanotubes.[43] They found the Kr capacity of the inner channels was $\sim 2.6 \times 10^{-3}$ mol/g at Kr partial pressure of 10 Pa at 77.3 K, twice the capacity of graphite at the same conditions. Jalili studied Kr and Xe adsorption on SWCNTs using Molecular Dynamic (MD) simulations.[44] They also found the advanced adsorption capacity of SWCNTs was attributable to the inner channels of the opened CNTs. Xe has a heat of adsorption twice as large as that of Kr, facilitating selective adsorption of Xe on SWCNTs. SWCNTs may present a cost-effective option to remove Xe in preliminary adsorption steps if low-cost production techniques can be developed.[36]

Zeolites are crystalline porous aluminosilicates composed of corner-sharing TO_4 (T=Al, Si, P) tetrahedrals with cages, cavities, and channels[45] that have been used to separate Kr/Xe mixtures. Zeolites are considered superior adsorbents to activated carbon since zeolites do not combust with

O_2 or NO_X. They are also chemically and physically stable at high temperatures. High surface areas allow greater storage capacities, and it is possible to fine-tune the adsorbate-adsorbent interactions to optimize selectivity. Zeolites are cost-effective adsorbents of noble gases since the adsorption can take place near ambient temperatures and moderate pressures. The size, shape, and structure of the zeolite pore, the cationic charge density, polarizability, and permanent polarity of the adsorbate molecules affect the selectivity and uptake of a gas on a zeolite. Smaller pores with a larger cation charge density will more selectively adsorb larger and more polar molecules.[47] Table 8.2 shows the Kr adsorption capacity of several zeolites as compared to activated carbon adsorbents.

Independent research groups have studied the capacity of zeolites to act as noble gas adsorbents. Jameson *et al.* conducted nuclear magnetic resonance (NMR) studies to investigate the competitive adsorption of a Kr–Xe mixture using the zeolite NaA.[49] The team confirmed that after a measure of time, Kr and Xe were adsorbed in NaA zeolite and formed Xe_nKr_m clusters. In a different study, NaA and NaX zeolites selectively adsorb Xe over Kr with selectivities of approximately 4–6. However, these zeolites showed

Table 8.2. Krypton capacities of various adsorbents. (Taken from Ref. 36).

Sorbent	Surface Area (m^2g^{-1})	Pressure (Pa)	Capacity of adsorbent (mol g^{-1})	Temp. (K)	Ref.
Natural mordenite[a]	217	25	6.50E−07	273	[40]
Cu-Mordenite[a]	—	20	3.50E−07	273	[40]
Hydrogen Mordenite[a]	295	15	4.00E−05	273	[40]
Silver Mordenite [a]	109	15	2.30E−06	273	[41]
MS-5A	415	30	2.70E−07	273	[40]
Activated Charcoal [b]	663	20	1.10E−06	273	[40]
Ambersoeb 572[b]	1100	30	2.20E−06	273	[40]
Ambersoeb 563[b]	—	40	2.00E−06	273	[40]
Molsievon X2M4/6[b]	—	20	1.70E−07	273	[40]
Molsievon 3A[b]	—	1000	2.00E−06	273	[40]
Merck[b]	—	10	5.00E−07	273	[40]
AUK microporous carbon[b]	—	2981	3.20E−03	177.7	[36]
SWCNT[b]	371.5	10	2.60E−03		[42]

[a]zeolite and [b]carbon.

low adsorption capacities (20–30 wt % at 100 kPa and 25°C).[47,50] Munakata *et al.* employed the zeolite molecular sieve 5A to adsorb Kr, Xe, N_2, and binary component systems (Kr-Xe, Kr-N_2).[38] The adsorption isotherms for molecular sieve 5A were correlated by the Langmuir model. In this study, Xe and large amounts of N_2 competitively inhibited the adsorption of Kr from the mixture of all three gases. Ianovsk and Munakata *et al.* synthesized the H-mordenite adsorbent from a natural mordenite by ion-exchange with an HCl solution.[41] The zeolite was found to adsorb Kr almost as well as activated charcoal. This research group also found that Xe exhibits an unusually strong interaction with silver-exchanged X and Y zeolites, compared with their sodium-exchanged counterpart.[42] In Y zeolites, the isosteric heats of adsorption of Xe on Ag-exchanged and Na-exchanged materials were 31.0 and 18.5 kJ/mol, respectively.[51] The authors believed that the ion-exchange treatment may have changed the electrostatic potential on the surface of the natural mordenite, greatly enhancing the surface affinity to noble gases. Nguyen *et al.* showed Ag^+ is the strongest binding site for Xe both in gas phase and on the chabazite surface with binding energies of 73.9 and 14.5 kJ/mol.[52] Although Ag^+ appears necessary for zeolites to bind Xe, the nature of this unusually strong binding is controversial. Recently, SAPO-34, a chabazite zeolite with pore size of 0.38 nm has been reported (in membrane form) as a promising material to effectively separate Kr from Xe.[53,54]

Metal-organic frameworks (MOFs)[55–72] have been identified as promising materials for noble gas separation. MOFs can store and separate Kr and Xe at ambient pressures, but typically they exhibit low adsorption capacity. Further research has uncovered MOFs with greater adsorption capacity, and additional research is expected to result in MOFs with increasingly high adsorption capacity. MOF adsorbents, such as IRMOF-1, HKUST-1, MOF-505, Ni-DOBDC, Co-formate, SBMOF-1, CC3, FMOF-Cu, and ZIF-8, have shown good potential for separating Kr and Xe from air. Mueller *et al.* investigated IRMOF-1, a MOF with tetranuclear $Zn_4O(BDC)_3$-based secondary building units (SBUs) to form channels with dimensions of ~1.5 nm.[64] The team was the first to investigate PSA performance for Xe/Kr (and Ar) at room temperature. The results showed strong preferential adsorptions for gases with higher polarizability (Xe > Kr > Ar). The simulated selectivity was only 3.7 for Ryan *et al.*,[57] and thus, IRMOF-1 does not make a strong candidate for industrial Xe/Kr separation.

HKUST-1 is a Cu-BTC MOF with cavities sized 1.3, 1.1, and 0.5 nm. At low pressures, HKUST-1 showed relatively high Xe selectivity (8.1 at 0.1 mPa), but at higher pressures, selectivity dropped off significantly, likely due to the presence of larger pores.[57] This demonstrated that MOFs may have uniformly small pore sizes for selectivity to persist at higher pressures.

MOF-505 is a 3-periodic MOF with copper paddlewheel SBUs and pore diameters of 0.48, 0.71, and 0.95 nm. Bae *et al.* measured Xe/Kr adsorption selectivity of 9–10.[65] The authors attribute the higher selectivity to the presence of smaller pore diameters compared to those of HKUST-1.

M-DOBDC (M-MOF-74) (M = Mg, Mn, Fe, Co, Ni, Cu, Zn; $dobdc^{4-}$ = 2,5-dioxido-1,4-benzenedicarboxylate) contains regular, hexagonal unidimensional pores sized at 1.1 nm.[64] At room temperature and 0.1 MPa, Ni-DOBDC was shown to have 12% higher selectivity than benchmark activated carbon.[66] Ni-DOBDC was also shown to have an Xe/Kr uptake ratio of 5–6 for most of the pressure range, compared to 3.8 for charcoal. At 0.1 MPa, the uptake rate of Ni-DOBCD was twice that of MOF-5.[66] Fernandez *et al.* found that Ni-DOBDC could separate 400 ppm Xe from 40 ppm Kr in air with a Xe/Kr selectivity of 7.3, making DOBDC a potential adsorbent for use in processing radioactive waste off-gases.[67]

Li *et al.* and Xiong *et al.* investigated the Xe/Kr adsorption selectivity of Co-formate, a flexible zinc tetrazolate framework with effective pore size of ~0.5–0.6 nm.[68,69] Xiong *et al.* measured the highest adsorption selectivity of 15.5 in a PSA process.[69] The pore size matches well with Xe, and along with the high adsorption enthalpy for Xe and Xe's high polarizability, Co-formate may be a good candidate for Xe-Kr separation applications.

SBMOF-1 is a calcium-based MOF, also known as CaSDB, SDB=4, 4-sulfonyldibenzoate.[70] Banerjee *et al.* performed computational simulations of SBMOF-1 and predicted that SBMOF-1 would have significantly greater selectivity than all other known MOFs in conditions similar to those encountered during nuclear fuel reprocessing. The prediction was experimentally confirmed.[71] They found a simulated selectivity 70.6 and largest included sphere diameter of 5.1 Å. They showed SBMOF has a high Xe adsorption capacity and a remarkable Xe/Kr selectivity of ~16 at 55 K. They calculated the potential energy of a Xe atom adsorbed in the pore (surface 1: $-32\,kJ\,mol^{-1}$; surface 2: $-15\,kJ\,mol^{-1}$). Using single-crystal diffraction, the authors found that the adsorption selectivity may be attributed to the pore geometry with polar –OH groups, which served as adsorption sites for polarizable Xe. The capacity and adsorption selectivity of SBMOF-1 are comparable to Co-formate, which makes SBMOF-1 one of the leading Xe-selective MOF adsorbents.

Pd-MOF is a MOF with sodalite structure that contains Pd^{2+} bonded to 2-hydroxypyrimidinolate linkers. The Pd atoms are incorporated in a square planar. This MOF has pore sizes of ~0.22, 0.49, and 0.58 nm.[57] Ryan *et al.* investigated Pd-MOF to test MOFs with even smaller pores.[57] The authors found that Xe/Kr selectivity was ~19 for the entire pressure range from 0.01 to 3 MPa. But Pd-MOF has remarkably lower adsorption capacities of both Xe and Kr than the other MOFs. It had the lowest void fraction (0.348) among all investigated MOFs.

All the above MOFs are Xe-selective. FMOF-Cu is a MOF with a pore size of 0.51 nm that selectively adsorbs Kr at temperatures below 0°C. FMOF-Cu shows a Kr/Xe selectivity of ~36 at 0.1 MPa and −70°C.[67] The pore size of FMOF-Cu is larger than the kinetic diameters of both Xe and Kr. However, the MOF is Kr selective because the connecting windows are comparably sized to Kr and smaller than Xe. This restricts the diffusion of Xe into the cavities. This property makes FMOF-Cu one excellent Kr-selective material candidate.

Porous organic cages (POC) are covalently bonded organic cages that assemble into crystalline microporous structures.[73,74] The porosity is intrinsic to the structure, rather than formed by non-covalent nonporous subunits. POCs are solution-processable and may achieve the atomic-scale pore-size control of MOFs. POCs have been reported with surface areas greater than 1500 $m^2\ g^{-1}$ and been shown to separate molecules based on size and shape.[75,76] The accessible surface area of POCs results from the interconnection of porous structures and empty spaces around the cages due to inefficient packing, also called extrinsic porosity. Chen *et al.* studied the adsorption properties of CC3, a porous organic cage.[77] The molecular structure of CC3 consists of repeating organic building blocks that form three-dimensional cages. After computer simulations, the authors conducted experiments and found that CC3 has excellent performance in the solid state for Kr/Xe gas separation. At the pressure range of $2 \times 10^{-3} - 5 \times 10^{-1}$ bar, the capacity for Xe was 10–30 times higher than the uptake for Kr.

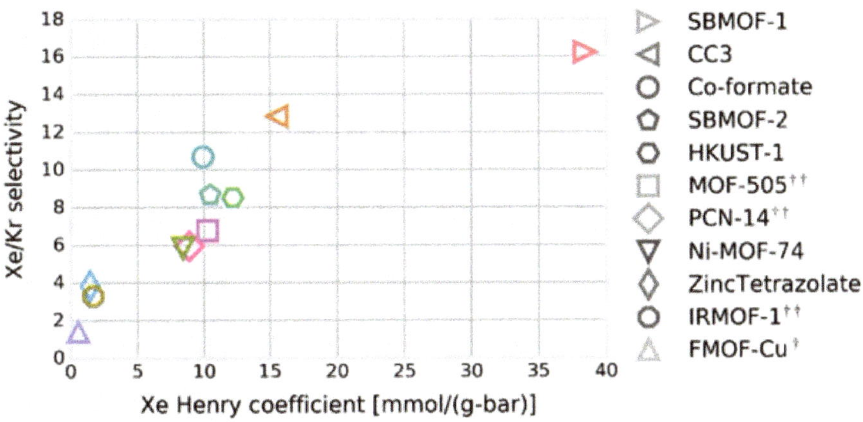

Figure 8.1. Xe/Kr adsorption selectivity as a function of Xe Henry coefficients of several MOFs and POCs. (Taken from Ref. 71).

Overall, MOF and POC adsorbents have demonstrated great potential to separate Kr and Xe. Figure 8.1 shows the Xe/Kr adsorption capacities as a function of Xe Henry coefficients for several MOFs and POCs.[71] SBMOF-1 and CC3 are the current leading Xe-selective materials. FMOF-Cu is the current leading Kr-selective material.

8.1.6. *ZIF-8 membranes for Kr/Xe separation*

To our best knowledge, the only report on any MOF membrane composition able to effectively separate Kr/Xe gas mixtures has been documented by Carreon research group.[72] We reported ZIF-8 membranes for Kr/Xe separation.[72] This particular MOF was chosen mainly based on its potential to molecularly sieve Kr over Xe. Based on the kinetic diameter of Kr (~0.36 nm) and Xe (0.41 nm), and the effective aperture size of ZIF-8 in the range of 0.4–0.42 nm,[78] ZIF-8 is an ideal candidate to molecular sieve Kr over Xe. In principle, Kr molecules would diffuse rapidly through the pores, while Xe, at most, will diffuse slowly, meaning that high Kr selectivities could be potentially achieved, based on differences in molecular diffusion.

ZIF-8 membranes were grown solvothermally on porous alumina tubes (Inopor GmbH, inside diameter of 0.7 cm and outside diameter of 1.1 cm, and are asymmetric within the inner layer, which has a pore size of 100 nm).

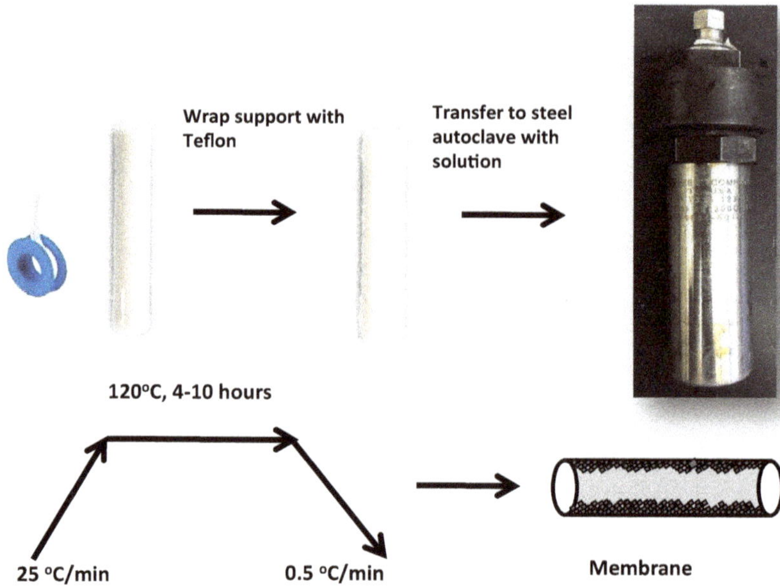

Figure 8.2. General synthesis approach for ZIF-8 membranes.

The outer surface of the support was wrapped with Teflon tape to prevent membrane growth on the outside surface. The support and the membrane gel were then placed in a stainless-steel autoclave (4713 General Purpose Pressure Vessel, 45 mL, Parr Instrument) and solvothermally treated in a conventional oven at 120°C for 4–10 hours. The gel covered the supports that were placed vertically in the autoclave. The general synthesis process for ZIF-8 membrane is shown in Figure 8.2.

After the autoclave was cooled down, the support was taken out, the Teflon tape was removed, and the support with the grown membrane was washed with methanol. For activation, the resultant 1-layered membrane was immersed in 25 mL of methanol for 1 day. A second and third layer (if needed) were applied following the same procedure described above. The resultant membranes were stored at ~80°C before gas permeation testing. Table 8.3 shows the detailed synthesis conditions for the ZIF-8 membranes.

The synthesized ZIF-8 membranes were used to separate premixed 92:8 Kr/Xe mixtures.[72] This Kr/Xe ratio is close to the typical ratio of these two gases in air. The feed pressure was 223 kPa, and the pressure in the

Table 8.3. General synthesis conditions of ZIF-8 membranes.

Membrane ID	Number of layers	General synthesis conditions
1A	2	1st layer: 120°C for 6 hours 2nd layer: 120°C for 6 hours
1B	2	1st layer: 120°C for 6 hours 2nd layer: 120°C for 6 hours
2A	2	1st layer: 120°C for 10 hours 2nd layer: 120°C for 4 hours
2B	2	1st layer: 120°C for 10 hours 2nd layer: 120°C for 4 hours
3	3	1st layer: 120°C for 10 hours 2nd layer: 120°C for 4 hours 3rd layer: 120°C for 4 hours

permeate side was 85 kPa. The separation results at room temperature for these membranes are shown in Table 8.5. Membranes 1A and 1B were prepared independently, using the same gel composition and same solvothermal synthesis conditions. Specifically, these two layered membranes were synthesized by growing both layers at 120°C for 6 hours. Kr permeances ranged from 0.17×10^{-8} mol/m^2 s Pa to 0.33×10^{-8} mol/m^2 s Pa. The permeances in units of Barrer ranged from ~14 to 104. For all studied membranes, the molar concentration of Kr in the permeate side ranged from 97.9% to 99.1%. Kr/Xe separation selectivities ranged from 5.9 to 10.8. The separation index π (which is defined as π = Kr permeance × (selectivity − 1) × permeate pressure) was calculated to assess membrane reproducibility. This index (π) has been used as a reliable parameter to predict porous crystalline membrane reproducibility. The separation index was $\pi = 13.6 \times 10^{-4}$ and 13.9×10^{-4} for membranes 1A and 1B, respectively, which indicated very good reproducibility. Two other membranes were synthesized, modifying only the solvothermal synthesis conditions with the hope of improving membrane separation performance. These two layered membranes prepared independently (2A and 2B) were synthesized by growing the first layer at 120°C for 10 hours (to promote better crystal intergrowth). In principle, longer solvothermal synthesis times may help particles to coalesce by diffusion leading to better particle intergrowth.

The second layer was grown at 120°C for only 4 hours, to limit crystal growth and obtain a thinner membrane. Membranes 2A and 2B displayed Kr permeances in the range from 1.3×10^{-8} mol/m^2 s Pa to 1.7×10^{-8} mol/m^2 s Pa, and Kr/Xe separation selectivities ranged from 12.3 to 16.1. Separation index values for membranes 2A and 2B were $\pi = 162 \times 10^{-4}$ and $\pi = 164 \times 10^{-4}$, respectively, again confirming membrane reproducibility. Membrane 3 was prepared by adding an extra ZIF-8 layer (to a membrane prepared similarly to membrane 2A or 2B) to form a three-layered membrane. The third layer was applied at 120°C for 4 hours. Adding this third layer was detrimental to the separation performance. Both Kr permeance and separation selectivities decreased, compared to the two layered membranes (2A and 2B). The decrease in permeance is attributed to an increase in membrane thickness. The reduction in separation selectivity may be associated with an increase in the concentration of defects and/or nonselective pore pathways. Defect formation and nonselective pathways are typical and distinctive features of any microporous crystalline membrane (zeolite, and MOF). These defects develop during membrane preparation, leading to grain-boundary formation, heterogeneous crystal growth, and limited crystal packing, among others. It is important to mention that we needed to add at least two layers to form continuous and reproducible membranes. Single-layered membranes were defective/discontinuous. The separation performance of these ZIF-8 membranes is summarized in Table 8.4.

Table 8.4. Kr/Xe Separation performance over ZIF-8 membranes (Taken from Ref. 72).

Membrane ID[a]	Kr permeance (mol/m^2 · s · Pa)[b]	Separation selectivity (α)	Separation index (π)[c]
1A	0.33×10^{-8} (9.9)	5.9	13.6×10^{-4}
1B	0.17×10^{-8} (5.1)	10.8	13.9×10^{-4}
2A	1.7×10^{-8} (50.8)	12.3	162×10^{-4}
2B	1.3×10^{-8} (38.8)	16.1	164×10^{-4}
3	0.5×10^{-8} (14.9)	7.9	28.9×10^{-4}

Molar gas mixture composition: 92:8 Kr/Xe. Transmembrane pressure 138 kPa. [a]1A,AB, 2A,2B are two layer membranes. 3 is three-layer membrane. [b]Numbers in parentheses indicate Gas permeation units (GPU). [c]π = Kr permeance × (selectivity-1) × permeate pressure.

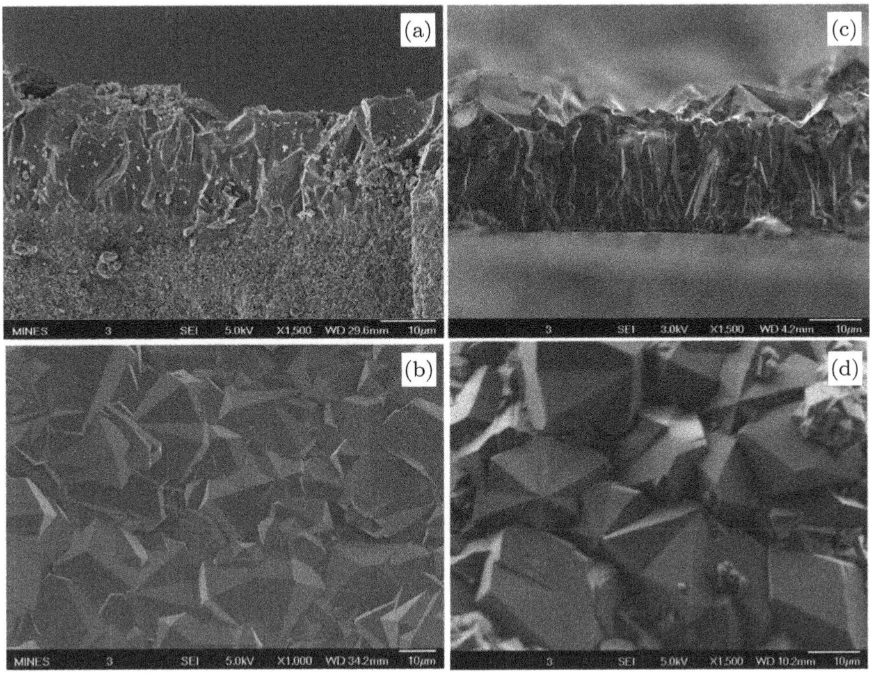

Figure 8.3. SEM images of ZIF-8 membranes: (top row) representative cross section and (bottom row) top view for membrane A1 (panels (a) and (b), respectively) and membrane 2B (panels (c) and (d), respectively). (Reprinted from Ref. 72 with permission from American Chemical Society.)

Figure 8.3 shows representative SEM images of membranes 1A and 2B.[72] Figure 8.3a shows the cross-sectional SEM image of membrane 1A, displaying a dense membrane layer with a thickness of $\sim 23 \pm 0.8\,\mu$m. Membrane 2B (Figure 8.3c) shows a slightly thinner layer of $\sim 20 \pm 1.2\,\mu$m. These results are consistent with the separation data shown in Table 8.5, which indicate higher Kr permeance for the thinner membrane 2B. The top-view SEM image for membranes 1A and 2B (Figures 8.3b and 8.3d, respectively) show well-intergrown and interconnected micrometer-range ZIF-8 crystals. A distinctive morphological feature of the most selective membrane (2B) is the presence of bipyramidal-shaped crystals. This morphology may develop as a result of the different solvothermal history of the membranes. Although we do not fully understand the potential (if any) role of this morphology in the separation performance, it is well-known that ZIF-8 polycrystalline membrane performance is highly dependent on microstructure.[79]

To better understand the separation mechanisms of the ZIF-8 membranes, adsorption isotherms, isosteric heats of adsorption, and column breakthrough experiments were collected for Kr and Xe. Figure 8.4a shows the adsorption isotherms for Kr and Xe measured at 298 and 278 K for ZIF-8 crystals. Single-component adsorption isotherms on ZIF-8 at room temperature indicate a Xe uptake of 1.70 mmol/g and a Kr uptake of 0.54 mmol/g.

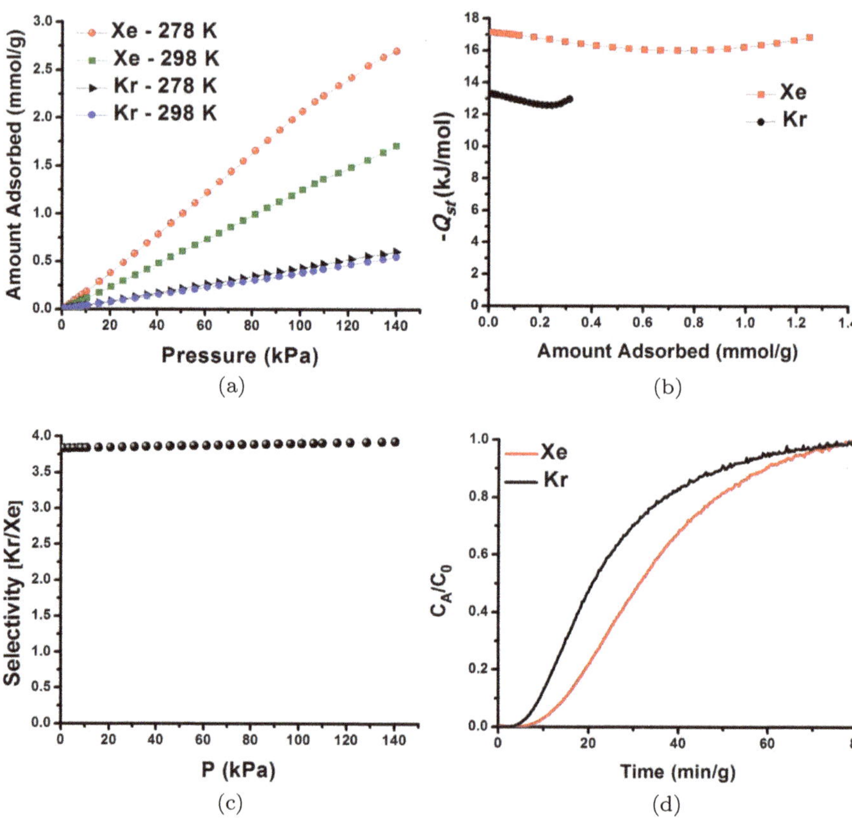

Figure 8.4. (a) Single-component Xe and Kr adsorption isotherms for ZIF-8 collected at 298 and 278 K, (b) Xe and Kr isosteric heats of adsorption (Qst) for ZIF-8 calculated by the virial method, (c) IAST calculated selectivity for 92:8 Kr/Xe gas mixture at 298 K for ZIF-8, and (d) column breakthrough experiment for 92:8 Kr/Xe gas mixture at 298 K and 140 kPa for ZIF-8. (Reprinted from Ref. 72 with permission from American Chemical Society.)

The overall Xe/Kr capacity at 140 kPa is lower than that of previously disclosed MOFs.[65,77,80–82] However, the trend is consistent, i.e., preferential adsorption of Xe over Kr. ZIF-8 preferentially adsorbed ~3.1 times more Xe than Kr. The leading sorbent material for Xe/Kr separation is the porous organic cage CC3, which was shown to separate 400 ppm Xe from 40 ppm Kr with a Xe/Kr selectivity of ~20.4.[77] The high adsorption selectivity of CC3 to Xe was attributed to a precise size match between Xe and the organic cage cavity. CC3 in membrane form has displayed molecular sieving properties for industrially relevant gas molecules, including CO_2, CH_4, N_2, H_2, and O_2.[83]

The isosteric heats of adsorption (Qst) of Xe and Kr gases were calculated by the virial method, using the experimental single adsorption isotherms collected at 298 and 278 K for both gases. ZIF-8 exhibits Qst values for Xe and Kr gases of −17 and −13 kJ/mol, respectively, which is in agreement with previously reported Qst values for Xe and Kr over ZIF-8.[84] These values remained almost steady by increasing the pressure up to 140 kPa for both gases, and confirm the preferential Xe adsorption over Kr (Figure 8.4b). This is attributed to the fact that Xe is a more polarizable molecule than Kr. The higher Xe polarizability (Xe, 26.85–28.7 atomic units; Kr, 16.44–18.0 atomic units)[85] leads to stronger van der Waals interactions with the −C=C– in the imidazole ligand of the ZIF-8, which is the only known polarizable group. The metal sites in the ZIF-8 are not accessible for any gas molecules, because of the steric hindrance by the surrounding ligands. *In situ* experiments with Xe and Kr confirm that these gas molecules have four adsorption sites: (i) the first preferential adsorption site is −C=C– of the imidazole group, (ii) the center of 6-membered window, (iii) within the 4-membered windows, and (iv) at higher pressure (between 93 to 400 mbar), the center of the pore is occupied with a gate opening mechanism.[86]

Ideal adsorbed solution theory (IAST) was used to predict the selectivity of Xe over a Kr binary mixture based on the experimental single adsorption isotherms collected at 298 K for Xe and Kr gases. The Kr selectivity of ZIF-8 crystals at 298 K for a 92:9 Kr/Xe binary gas mixture was ~4 (Figure 8.4c). This Kr/Xe selectivity for the 92:8 Kr/Xe feed mixture reflects the uptake of the Kr at 92% of the total pressure (1.4 bar) with the uptake of Xe at 8% of the total pressure (1.4 bar), using the Kr and Xe isotherms collected at 298 K.

These results indicate that, under the separation conditions employed, ZIF-8 can adsorb more Kr than Xe, which is reasonable, based on the high Kr concentration (Kr/Xe: 92:8). To determine the relative diffusivity differences between Kr and Xe over ZIF-8 crystals, column breakthrough experiments were conducted at 298 K and 140 kPa, for a 92:8 Kr/Xe gas mixture over ZIF-8 (Figure 8.4d). The breakthrough time was ~2.6 min for Kr and ~6 min for Xe for 1 g of sample at a flow rate of 5 mL/min, indicating a higher diffusivity of Kr over Xe. Kr/Xe selectivity were also calculated from the breakthrough experiments at room temperature by calculating the Xe and Kr adsorption capacities, as shown in our previous report.[80] The calculated Kr/Xe selectivity was found to be 5.1, which is in good agreement with the value obtained by IAST calculation. These results are in agreement with the lower capacity of Kr and the affinity of ZIF-8 to diffuse out from the column faster, while Xe has higher affinity and capacity for ZIF-8 breakthrough later. Breakthrough column experiments suggest that differences in diffusivity favor the separation of Kr over Xe in the gas mixture.

Three separation mechanisms played an important role for the separation of Kr/Xe gas mixtures over ZIF-8 membranes — molecular sieving, diffusivity differences, and preferential adsorption. The separation results showing Kr selective membranes, and the fact that the effective aperture size of ZIF-8 in the range of 0.4–0.42 nm lies approximately between the kinetic diameter of Kr (~0.36 nm) and Xe (~0.41 nm) suggests molecular sieving properties of ZIF-8 membranes for this binary gas mixture. Because of the ZIF-8 flexibility associated with the effective pore size range of 0.4–0.42 nm, sharp molecular sieving effect may be limited. However, based on the observed separation, adsorption, and breakthrough data, molecular sieving plays an important role as separation mechanism. Breakthrough experiments confirmed higher diffusivity of Kr over Xe, favoring again Kr selective membranes. Although adsorption isotherms and isosteric heat of adsorptions over ZIF-8 crystals suggest that Xe adsorbs more strongly than Kr, IAST predicted that at the prevailing gas separation conditions (Kr/Xe: 92:8 feed composition), Kr/Xe adsorption selectivity was ~4, which also favors Kr selective membranes.

Our group compared three different types of microporous crystalline membranes for Kr/Xe separation, namely SAPO-34 (zeolite), ZIF-8 (MOF), and AlPO-18 (alumino phosphate molecular sieve).[87] The separation

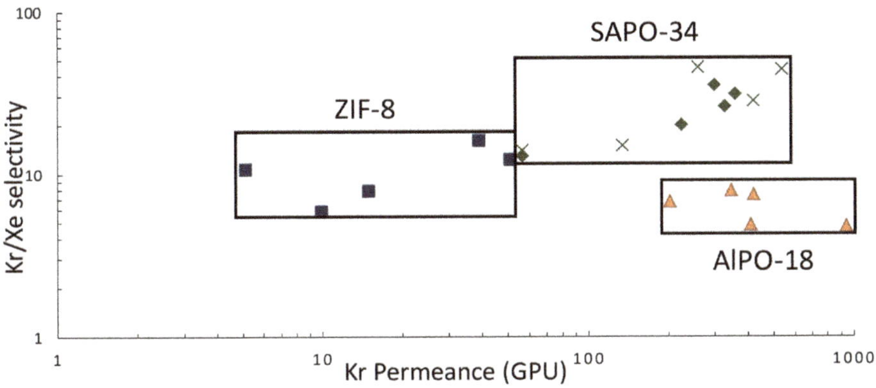

Figure 8.5. Kr/Xe separation selectivity vs Kr permeance over AlPO-18, SAPO-34, and ZIF-8 membranes for a Kr-rich feed gas mixture (molar gas mixture composed of 92:8 Kr/Xe was employed for AlPO-18 (yellow triangles), SAPO-34 (green diamonds) and ZIF-8 (blue rectangles), and 9:1 Kr/Xe molar gas mixture for SAPO-34 (green crosses)) (Reprinted from Ref. 87 with permission from American Chemical Society.)

performance of our previously reported SAPO-34[53] and ZIF-8[72] for Kr/Xe separation vs AlPO-18[87] membranes is shown in Figure 8.5. These results allow a direct comparison between these three distinctive types of microporous crystalline membranes since all these membranes were grown on the same type of porous supports and were evaluated under similar separation conditions.

SAPO-34 displayed the highest average Kr/Xe separation selectivity. Two main factors contributed to this enhanced observed selectivity. The first factor is related to the intrinsic molecular sieve property of SAPO-34 imparted by its rigid pore size of 0.38 nm. The second factor is related to the lowest (of the three membrane compositions) Xe/Kr adsorption selectivity which translates into an attenuated competitive adsorption effect, and therefore resulting into higher Kr/Xe separation selectivities.[87] AlPO-18 membranes displayed the highest average Kr permeances, which can be explained by membrane thickness. AlPO-18 membranes displayed the thinnest layers of all studied membranes. In addition, the highest Kr permeances observed in AlPO-18 membranes were associated to a higher concentration of non-selective pathways.[87] As compared to AlPO-18 and SAPO-34, ZIF-8 membranes were thicker (~one order of magnitude) resulting in the lowest Kr permeances. The moderate Kr/Xe separation

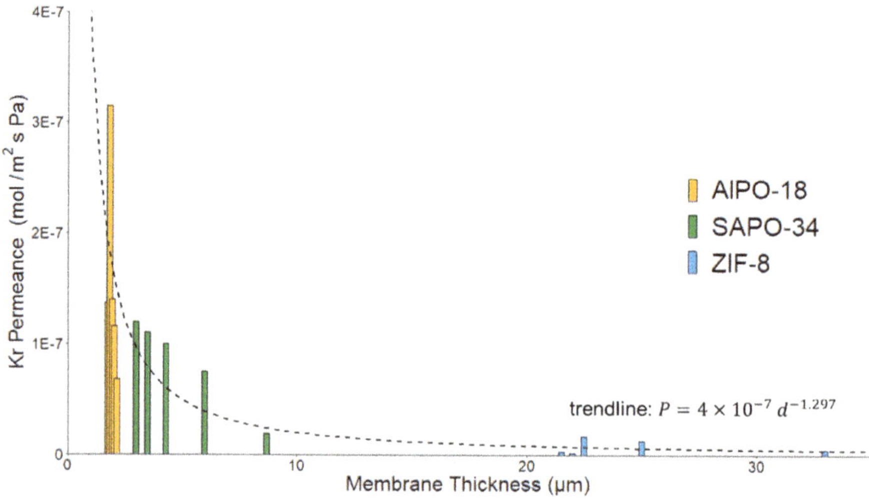

Figure 8.6. Kr permeance for SAPO-34, ZIF-8 and AlPO-18 membranes as a function of membrane thickness over Kr-rich feed gas mixture. Dashed line reflects the fitting equation. (Reprinted from Ref. 87 with permission from American Chemical Society.)

selectivities observed over ZIF-8 membranes were associated to the flexible nature of its pore structure, limiting its ability to display sharp molecular sieving, and potentially to the presence of defects.[87]

The Kr permeances as a function of membrane thickness for all three membrane compositions is shown in Figure 8.6. The thinner membranes (AlPO-18) displayed Kr permeances about 10–15 times higher than the thickest membranes (ZIF-8). As illustrated by the dashed line on Figure 8.6, the relation between thickness and permeance decreases according to a power function, and follows the equation: $P = 4 \times 10^{-7} d^{-1.297}$ where d is membrane thickness (μm) and P is Kr permeance (mol/m$^2 \cdot$ s $\cdot$ Pa).[87] This correlation may be useful to predict the Kr permeance of other porous crystal compositions with similar micropore sizes of the studied crystals. Therefore, if ZIF-8 membranes are prepared with thicknesses below ~5 μm, they could improve by at least one order of magnitude its Kr permeance and potentially compete with SAPO-34 and/or AlPO-18 membranes. Furthermore, recent molecular simulation studies on porous crystalline membranes predict that fluorinating the heterocycle of the imidazole linker in ZIF-8 may lead to a membrane selectivity of an order of magnitude higher than that of pure

ZIF-8 membrane.[88] These simulation insights are highly promising and need to be proven experimentally.

8.2. Air/Xe separation

8.2.1. *Introduction*

The use of Xenon becomes remarkable when its cost is considered. For instance, Xe costs about $262 for 11 liters at market rates.[89] A decrease in market price for high quality Xe would benefit the medical and commercial industries, as well as promote the development of new Xe applications. The high cost of Xe gas derives from the current energy- and capital-intensive cryogenic distillation separation process of air, and the low concentration of Xe in air, which is typically 0.086 ppm. Cryogenic distillation requires cooling air below the boiling points of nitrogen and oxygen (Nitrogen: −196°C, Oxygen: −183°C, Xenon: −108°C)[90] to separate the inert gas from the mixture. In a conventional cryogenic air separation plant, intensive capital and energy costs are expended in a series of process units that involve air compression, air purification, heat exchanging, distillation, and product compression.[91] Because this process is expensive, a limited amount of Xe is separated from air. The yearly world production of Xenon is about 80.7 metric tonnes (14 million liters) while Earth's atmosphere contains approximately 2 billion tonnes.[92]

8.2.2. *Adsorbents for Xenon separation from air*

Several adsorbents have been employed to separate Xe from air as a potentially cheaper approach as compared to cryogenic distillation. This technology (similar to the employed approach for the separation of Krypton from Xenon) relies on the adsorption capacity difference between Xenon and air components (oxygen and nitrogen) rather than on phase change differences. In most of these reports, carbon, zeolite and metal organic frameworks (MOFs) have been used to capture Xenon and other noble gases from air.[93–96] For example, Golden *et al.* demonstrated Li- and Ag-exchange zeolites as effective absorbents to recover Xe and Kr from an oxygen-containing gas derived from a liquid oxygen stream in a cryogenic air separation plant.[93] Bazan *et al.* studied the adsorption difference between gaseous oxygen, krypton, xenon and argon over zeolites, and activated carbons.[94] These authors found that Xe adsorbed preferentially over the other gases.[94] Parkes *et al.* screened several metal-organic frameworks with distinctive pore sizes and topologies for adsorbing noble gases in air.[107] The authors found that the best MOFs displayed Xe to nitrogen selectivity as high as 26.[107] Thallapally *et al.* demonstrated that a Ni-based MOF displays higher Xe to nitrogen adsorption selectivity than activated carbons.[96]

8.2.3. *ZIF-8 membranes for Xenon separation from air*

Although adsorbents for Xe separation from air have been reported, there is limited research on membrane technology for capturing Xe from air. Recently, Carreon group and collaborators documented the first example of any microporous crystalline membrane (ZIF-8) displaying separation ability for air/Xe mixtures.[97] In the next paragraphs, we describe the relevant results, and important scientific insights resulted from this study.

Similar to Kr/Xe separation, the rationale of employing ZIF-8 membranes for Xe separation from air is the ability of this MOF composition to potentially molecular sieve nitrogen and oxygen from xenon (size exclusion effect). The separation performance of ZIF-8 membranes prepared in this study was evaluated using premixed air/Xe gas mixtures with molar ratio of 9:1.

The separation data for three ZIF-8 membranes is shown in Table 8.5. Two membranes M1 and M2 were synthesized independently via solvothermal approach. The first layer of membranes M1 and M2 was prepared at

Table 8.5. Air/Xe separation performance over ZIF-8 membranes. (Taken from Ref. 97).

Membrane ID	Air permeance[a] $mol/m^2 \cdot s \cdot Pa$	Separation selectivity (α)	Separation index (π)
M1	3.10×10^{-8} (92.6)	11.3	2.7×10^{-2}
M2	3.94×10^{-8} (117.7)	12.4	3.8×10^{-2}
M3	3.94×10^{-8} (117.7)	7.4	2.1×10^{-2}

Molar gas mixture composition: 9:1 air/Xe; transmembrane pressure: 138 kPa; feed flow rate: 40 ml/min. [a]Numbers in parenthesis indicate Gas Permeation Units (GPU).

120°C for 10 hours. Extended solvothermal synthesis times allow better crystalline intergrowth. We applied two layers for these membranes since one-layer membranes were defective/discontinuous. The second layer of M1 and M2 was added over a shorter duration of three and four hours respectively at 120°C to limit crystal growth and form a continuous layer. M1 and M2 displayed permeances of 3.1 and 3.9 × 10^{-8} mol/m^2 s Pa and air/Xe separation selectivities of 11.3 and 12.4 respectively. Membrane M3 was prepared by adding an extra ZIF-8 layer to a membrane prepared with identical gel compositions and solvothermal synthesis conditions as those of M1 to form a 3-layer membrane. Separation selectivity decreased compared to those of the 2-layer membranes (M1 and M2). The reduction in separation selectivity may be associated with an increase in the concentration of defects and/or non-selective pore pathways.

The separation index π was used to assess membrane reproducibility. Membrane separation indexes ranged between 2.1 × 10^{-2} and 3.8 × 10^{-2} mol/m^2 s respectively, indicating good membrane reproducibility.

To learn about the role of adsorption of nitrogen and Xenon over ZIF-8, adsorption isotherms were collected for N_2 and Xe. In this study,[97] we employed pure nitrogen as a reasonable approximation of air composition. Figure 8.7 shows the single adsorption isotherms for N_2 and Xe measured at 10°C, and 25°C for ZIF-8 crystals. The isotherms indicate that the Xe/N_2 adsorption selectivity decreases as temperature decreases. Specifically, the

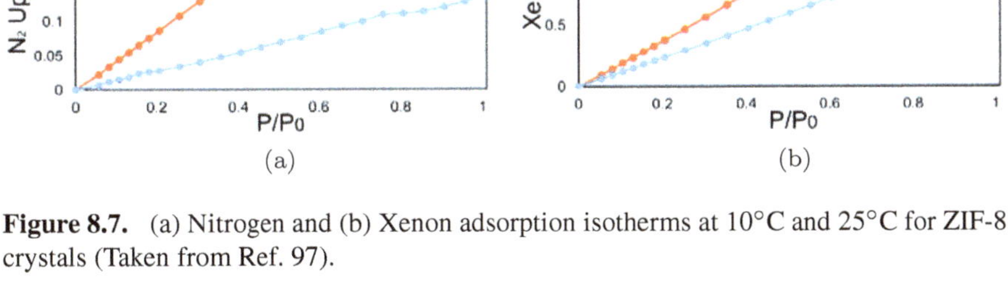

Figure 8.7. (a) Nitrogen and (b) Xenon adsorption isotherms at 10°C and 25°C for ZIF-8 crystals (Taken from Ref. 97).

Xe/N_2 adsorption selectivity at $P/Po = 1$, was 9.0 and 4.8 at 25°C and 10°C respectively.

The isosteric heats of adsorption (Qst) of Xe and N_2 gases were calculated by Virial method, using the experimental single adsorption isotherms collected at 298 K and 283 K for both gases. ZIF-8 exhibits Qst values for Xe and N_2 of ~ -17 and ~ -11 kJ/mol, respectively. These values indicate Xe preferential adsorption over N_2, a fact attributable to the greater polarizability of Xe molecules. The higher Xe static dipole polarizability (Xe: 25.297-27.42 atomic units; N2:7.26 atomic units 15) leads to stronger van der Waals interactions with the –C=C– polar ligand of ZIF-8. Therefore, stronger electrostatic interactions between the ZIF-8 surface and Xe lead to Xe preferential adsorption. The higher Xe adsorption over ZIF-8 indicates that adsorption between Xe and N_2 is a strong competitive separation mechanism.

We postulated that molecular sieving and differences in diffusivities were the two main separation mechanisms which contributed to the observed air over Xe selective ZIF-8 membranes.[97] The effective aperture size of ZIF-8 (~0.4–0.42 nm), lying between the kinetic diameters of nitrogen and oxygen (~0.35–0.36 nm) and Xe (~0.41 nm), suggests potential molecular sieving properties of ZIF-8 membranes for this gas mixture. Estimation of Fickian diffusivities indicate that air diffuses faster than Xe promoting the separation via differences in diffusivities. On the other hand, the observed Xe/air preferential adsorption selectivity (Figure 8.7) competes strongly against molecular sieving and diffusivity differences. The structural framework flexibility of ZIF-8 leads to Xe permeation. This "gate opening functionality" can change the ZIF effective pore aperture helping to promote the separation via differences in diffusivities. ZIF-8 pore size is close enough to potentially promote molecular sieving too.

8.3. Concluding remarks

If prepared in membrane form, several MOF compositions may be appealing to effectively separate Kr from Xe, a highly challenging molecular gas mixture. Some of these MOF compositions aim to exploit their uniform microporous pore windows (limiting pore apertures) to molecular sieve Kr from Xe, as well as its tunable surface chemistry to promote the separation by preferential adsorption. In terms of potential molecular sieving effect, zeolitic imidazolate frameworks (ZIFs) are in particular appealing, since some of them have effective pore sizes which lie between the kinetic diameters of Kr and Xe. However, one of the main technical challenges is to establish robust and reproducible synthetic protocols that could lead to the development of high quality MOF membranes. Up to date, and to our best knowledge there is only one successful example of a continuous MOF membrane for the separation of Kr/Xe gas mixtures. This membrane composition is the zeolite imidazolate framework-8, one of the most studied ZIF membranes for the separation of diverse molecular gas mixtures. The best ZIF-8 membranes separated Kr/Xe mixtures with average Kr permeances as high as 1.5×10^{-8} mol/m^2 s Pa and average separation selectivities of 14.2 for molar feed compositions corresponding to Kr/Xe ratio encountered typically in air.

ZIF-8 membranes have been effective in separating Air/Xe mixtures too. These membranes displayed air permeances as high as 3.9×10^{-8} mol/m^2 s Pa and separation selectivities as high as 12.4 for air/Xe feed molar compositions of 9:1. Separation mechanisms included molecular sieving, competitive adsorption, and diffusivity differences. The former two mechanisms were dominant leading to air selective membranes.

Although these membranes represent a promising alternative to the benchmark technology cryogenic distillation currently employed to separate xenon from krypton and air, mixture, further research efforts are needed to improve the separation performance of these membranes. In addition, the long term stability, separation performance in the presence of impurities, effect of transmembrane pressure, among other studies are needed to better assess the potential use of these membranes for these particular gas mixtures. Future studies also should focus on other suitable MOF compositions that could be reproducibly grown in membrane form.

References

1. Energy.gov. https://www.energy.gov/ne/articles/nuclear-power-most-reliable-energy-source-and-its-not-even-close
2. On-Site Storage of Nuclear Waste Used Nuclear Fuel and High-Level Radioactive Waste https://www.nei.org/Knowledge-Center/Nuclear-Statistics/On-Site-Storage-of-Nuclear-Waste.
3. Processing of Used Nuclear Fuel http://www.world-nuclear.org/information-library/nuclear-fuel-cycle/fuel-recycling/processing-of-used-nuclear-fuel.aspx.

4. Crawford, P. G. Zeolite membranes for the separation of krypton and xenon from spent nuclear fuel reprocessing off-gas (Doctoral dissertation, Georgia Institute of Technology) (2013).
5. Soelberg, N. & Watson, T. Deep Bed Iodine Sorbent Testing FY 2011 Report (No. INL/EXT-11-23191). Idaho National Laboratory (INL) (2011).
6. Banerjee, D., Cairns, A. J., Liu, J., Motkuri, R. K., Nune, S. K., Fernandez, C. A., Krishna, R., Strachan, D. M. & Thallapally, P. K. Potential of metal-organic frameworks for separation of xenon and krypton. *Accounts of Chemical Research* **48**(2), 211–219 (2014).
7. Abe, K., Hosaka, J., Iida, T., Ikeda, M., Kobayashi, K., Koshio, Y. & Nakajima, Y. Distillation of liquid xenon to remove krypton. *Astroparticle Physics* **31**(4), 290–296 (2009).
8. NCRP Report 44: Krypton-85 In The Atmosphere-Accumulation, Biological Significance, and Control Technology. Washington, DC, NCRP, 1975.
9. NCRP Report: Krypton-85 In The Atmosphere-With Specific Reference to the Public Health Significance of the Proposed Controlled Release at Three Mile Island. Washington, DC, NCRP, 1980.
10. Gombert, D., Ebert, W., Marra, J., Jubin, R. & Vienna, J. Global nuclear energy partnership waste treatment baseline. Papers, 235th ACS Nat. Meet., New Orleans, LA (The United States) (2008).
11. Boyum, B. M. An economic evaluation of the recovery of krypton and xenon from nuclear fuels reprocessing plants (Master's thesis, University of Arizona.) (1971).
12. Brunner, R., Haina, D., Landthaler, M., Waidelich, W. & Braun-Falco, O. Applications of laser light of low power density. Experimental and clinical investigations. In Therapeutic Photomedicine (pp. 111–116). Karger Publishers (1986).
13. Severinghaus, J. P., Grachev, A., Luz, B. & Caillon, N. A method for precise measurement of argon 40/36 and krypton/argon ratios in trapped air in polar ice with applications to past firn thickness and abrupt climate change in Greenland and at Siple Dome, Antarctica. *Geochimica et Cosmochimica Acta* **67**(3), 325–343 (2003).
14. Costello, S., Desmulliez, M. P. & McCracken, S. Review of test methods used for the measurement of hermeticity in packages containing small cavities. *IEEE Transactions on Components, Packaging and Manufacturing Technology* **2**(3), 430–438 (2012).
15. Sayre, E. D. Commercial Value of Used Nuclear Fuel Reprocessed with Elements Separated, Purified and Reduced to Metals.
16. Doke, T. A historical view on the R&D for liquid rare gas detectors. *Nuclear Instruments and Methods in Physics Research Section A: Accelerators, Spectrometers, Detectors and Associated Equipment* **327**(1), 113–118 (1993).
17. MPalow, J. E. U.S. Patent No. 5,712,650. Washington, DC: U.S. Patent and Trademark Office (1998).
18. Oesterle, L. J., Newman, S. M. & Shellhart, W. C. Rapid curing of bonding composite with a xenon plasma arc light. *American Journal of Orthodontics and Dentofacial Orthopedics* **119**(6), 610–616 (2001).
19. Richards, G. B. The application of electronic video techniques to infrared and ultraviolet examinations. *Journal of Forensic Science*, **22**(1) 53–60 (1977).
20. Lachmann, B., Armbruster, S., Schairer, W., Landstra, M., Trouwborst, A., van Daal, G. J., Kusuma, A. & Erdmann, W. Safety and efficacy of xenon in routine use as an inhalational anaesthetic. *The Lancet*, **335**(8703), 1413–1415 (1990).

21. Sanders, R. D., Franks, N. P. & Maze, M. Xenon: No stranger to anaesthesia. *British Journal of Anaesthesia* **91**(5), 709–717 (2003).
22. Wanner, A. U.S. Patent Application No. 11/509,809 (2006).
23. Li, J. R., Kuppler, R. J. & Zhou, H. C. Selective gas adsorption and separation in metal-organic frameworks. *Chemical Society Reviews* **38**(5), 1477–1504 (2009).
24. Soelberg, N. R., Garn, T. G., Greenhalgh, M. R., Law, J. D., Jubin, R., Strachan, D. M. & Thallapally, P. K. Radioactive iodine and krypton control for nuclear fuel reprocessing facilities. *Science and Technology of Nuclear Installations* (2013).
25. Yusa, H., Kikuchi, M., Tsuchiya, H., Kawaguchi, O. & Segawa, T. Application of cryogenic distillation to Krypton-85 recovery. *Nuclear Engineering and Design* **41**(3), 437–441 (1977).
26. Offutt, G. F. & Bendixsen, C. L. Rare Gas Recovery Facility at the Idaho Chemical Processing (No. IN–1221). Idaho Nuclear Corp., Idaho Falls (1969).
27. Pence, D. T. "Critical review of noble gas treatment systems," in *Proceedings of the 16th DOE Nuclear Air Cleaning Conference*, vol. 2 of CONF-80138, pp. 989–1013, The Harvard Air Cleaning Laboratory, Boston, Mass, USA (1980).
28. Aprile, E., Alfonsi, M., Arisaka, K., Arneodo, F., Balan, C., Baudis, L. & Brown, E. Dark matter results from 225 live days of XENON100 data. *Physical Review Letters* **109**(18), 181301 (2012).
29. Wang, Z., Bao, L., Hao, X. & Ju, Y. Design and construction of a cryogenic distillation device for removal of krypton for liquid xenon dark matter detectors. *Review of Scientific Instruments* **85**(1), 015116 (2014).
30. Aprile, E., *et al.* Removing krypton from xenon by cryogenic distillation to the ppq level. arXiv preprint arXiv:1612.04284 (2016).
31. Chen, L., *et al.* Separation of rare gases and chiral molecules by selective binding in porous organic cages. *Nature Materials* **13**(10), 954–960 (2014).
32. Zabaluev, Y. V. Management of radionuclides from reprocessing plant gaseous effluents. *IAEA Bull* **21**(1), 23–31 (1979).
33. Little, D. K., Eby, R. S., Norton, J. L., Patton, J. L., Schultz, R. M. & Varagona, J. M. Noble gas removal and concentration by combining fluorocarbon absorption and adsorption technologies (No. K/PS-5002; CONF-820833-8). Oak Ridge Gaseous Diffusion Plant, TN (USA) (1982).
34. IAEA, "Separation, storage and disposal krypton-85," Technical Report Series No. 199, International Atomic Energy Agency, Vienna, Austria (1980).
35. Substance: Dichlorodifluoromethane http://www.rsc.org/learn-chemistry/wiki/Substance:Dichlorodifluoromethane
36. Nandanwar, S. U., Coldsnow, K., Utgikar, V., Sabharwall, P. & Aston, D. E. (2016). Capture of harmful radioactive contaminants from off-gas stream using porous solid sorbents for clean environment — A review. *Chemical Engineering Journal* **306**, 369–381 (1982).
37. Potapov, S. V., Fomkin, A. A., Sinitsyn, V. A. & Shkolin, A. V. Krypton adsorption on microporous adsorbents at higher pressures. *Protection of Metals and Physical Chemistry of Surfaces* **46**(6), 639–643 (2010).
38. Munakata, K., Fukumatsu, T., Yamatsuki, S., Tanaka, K. & Nishikawa, M. Adsorption equilibria of krypton, xenon, nitrogen and their mixtures on molecular sieve 5A and activated charcoal. *Journal of Nuclear Science and Technology* **36**(9), 818–829 (1999).

39. Munakata, K., Yamatsuki, S., Tanaka, K. & Fukumatsu, T. Screening test of adsorbents for recovery of krypton. *Journal of Nuclear Science and Technology* **37**(1), 84–89 (2000).
40. Munakata, K., Shinozaki, T. & Okabe, H. Adsorption of krypton on adsorbents at cryogenic temperatures. *Journal of Power and Energy Systems* **2**(1), 171–177 (2008).
41. Ianovski, D., Munakata, K., Kanjo, S., Yokoyama, Y., Koga, A., Yamatsuki, S., Tanaka, K., Fukumatsu, T., Nishikawa, M. & Igarashi, Y. Adsorption of noble gases on H-mordenite. *Journal of Nuclear Science and Technology* **39**(11), 1213–1218 (2002).
42. Munakata, K., Kanjo, S., Yamatsuki, S., Koga, A. & Ianovski, D. Adsorption of noble gases on silver-mordenite. *Journal of Nuclear Science and Technology* **40**(9), 695–697 (2003).
43. Muris, M., Dufau, N., Bienfait, M., Dupont-Pavlovsky, N., Grillet, Y. & Palmari, J. P. Methane and krypton adsorption on single-walled carbon nanotubes. *Langmuir* **16**(17), 7019–7022 (2000).
44. Jalili, S. & Majidi, R. Study of Xe and Kr adsorption on open single-walled carbon nanotubes using molecular dynamics simulations. *Physica E: Low-dimensional Systems and Nanostructures* **39**(1), 166–170 (2007).
45. Breck, D. W. *Zeolite Molecular Sieves, Structure, Chemistry, and Use*, John Wiley & Sons, Inc., New York, NY (1974).
46. Ramsahye, N. A. & Bell, R. G. Cation mobility and the sorption of chloroform in zeolite NaY: Molecular dynamics study. *The Journal of Physical Chemistry B* **109**(10), 4738–4747 (2005).
47. Auerbach, S. M., Carrado, K. A. & Dutta, P. K. *Handbook of Zeolite Science and Technology*. CRC Press (2003).
48. Sircar, S. & Myers, A. L. Gas separation by zeolites. Handbook of Zeolite Science and Technology, 1063 (2003).
49. Jameson, C. J., Jameson, A. K. & Lim, H. M. Competitive adsorption of xenon and krypton in zeolite Na A: 129 Xe nuclear magnetic resonance studies and grand canonical Monte Carlo simulations. *The Journal of Chemical Physics* **107**(11), 4364–4372 (1997).
50. Izumi, J., Auerbach, S., Carrado, K. & Dutta, P. Handbook of Zeolite Science and Technology (2003).
51. Watermann, J. & Boddenberg, B. Isosteric heats of adsorption of xenon in silver-exchanged Y zeolites. *Zeolites* **13**(6), 427–429 (1993).
52. Nguyen, H. G., Konya, G., Eyring, E. M., Hunter, D. B. & Truong, T. N. Theoretical study on the interaction between xenon and positively charged silver clusters in gas phase and on the (001) chabazite surface. *The Journal of Physical Chemistry C*, **113**(29), 12818–12825 (2009).
53. Feng, X., Zong, Z., Elsaidi, S. K., Jasinski, J. B., Krishna, R., Thallapally, P. K. & Carreon, M. A. Kr/Xe Separation over a Chabazite Zeolite Membrane. *Journal of the American Chemical Society*, **138**(31), 9791–9794 (2016).
54. Hye Kwon, Y., Kiang, C., Benjamin, E., Crawford, P., Nair, S. & Bhave, R. Krypton-xenon separation properties of SAPO-34 zeolite materials and membranes. *AIChE Journal* **63**(2), 761–769 (2017).
55. Rowsell, J. L. & Yaghi, O. M. Metal-organic frameworks: A new class of porous materials. *Microporous and Mesoporous Materials* **73**(1), 3–14 (2004).

56. Férey, G. Hybrid porous solids: Past, present, future. *Chemical Society Reviews* **37**(1), 191–214 (2008).
57. Ryan, P., Farha, O. K., Broadbelt, L. J. & Snurr, R. Q. Computational screening of metal-organic frameworks for xenon/krypton separation. *AIChE Journal* **57**(7), 1759–1766 (2011).
58. Koh, K., Wong-Foy, A. G. & Matzger, A. J. A porous coordination copolymer with over 5000 m^2/g BET surface area. *Journal of the American Chemical Society* **131**(12), 4184–4185 (2009).
59. Sikora, B. J., Wilmer, C. E., Greenfield, M. L. & Snurr, R. Q. Thermodynamic analysis of Xe/Kr selectivity in over 137000 hypothetical metal-organic frameworks. *Chemical Science* **3**(7), 2217–2223 (2012).
60. Parkes, M. V., Staiger, C. L., Perry IV, J. J., Allendorf, M. D. & Greathouse, J. A. (2013). Screening metal-organic frameworks for selective noble gas adsorption in air: Effect of pore size and framework topology. *Physical Chemistry Chemical Physics* **15**(23), 9093–9106 (2012).
61. Hulvey, Z., Lawler, K. V., Qiao, Z., Zhou, J., Fairen-Jimenez, D., Snurr, R. Q., Ushakov, S. V., Navrotsky, A., Brown, C. M. & Forster, P. M. Noble gas adsorption in copper trimesate, HKUST-1: An experimental and computational study. *The Journal of Physical Chemistry C* **117**(39), 20116–20126 (2013).
62. Perry IV, J. J., Teich-McGoldrick, S. L., Meek, S. T., Greathouse, J. A., Haranczyk, M. & Allendorf, M.´D. Noble gas adsorption in metal-organic frameworks containing open metal sites. *The Journal of Physical Chemistry C* **118**(22), 11685–11698 (2014).
63. Heest, T., Teich-McGoldrick, S. L., Greathouse, J. A., Allendorf, M. D. & Sholl, D. S. Identification of metal-organic framework materials for adsorption separation of rare gases: Applicability of ideal adsorbed solution theory (IAST) and effects of inaccessible framework regions. *J. Phys. Chem. C* **116**, 13183 (2012).
64. Mueller, U., Schubert, M., Teich, F., Puetter, H., Schierle-Arndt, K. & Pastre, J. Metal-organic frameworks — prospective industrial applications. *Journal of Materials Chemistry* **16**(7), 626–636 (2006).
65. Bae, Y. S., Hauser, B. G., Colón, Y. J., Hupp, J. T., Farha, O. K. & Snurr, R. Q. High xenon/krypton selectivity in a metal-organic framework with small pores and strong adsorption sites. *Microporous and Mesoporous Materials* **169**, 176–179 (2013).
66. Liu, J., Thallapally, P. K. & Strachan, D. Metal-organic frameworks for removal of Xe and Kr from nuclear fuel reprocessing plants. *Langmuir* **28**(31), 11584–11589 (2012).
67. Fernandez, C. A., Liu, J., Thallapally, P. K. & Strachan, D. M. Switching Kr/Xe selectivity with temperature in a metal-organic framework. *Journal of the American Chemical Society* **134**(22), 9046–9049 (2012).
68. Li, K., Olson, D. H., Lee, J. Y., Bi, W., Wu, K., Yuen, T., Xu, Q. & Li, J. Multifunctional microporous MOFs exhibiting gas/hydrocarbon adsorption selectivity, separation capability and three-dimensional magnetic ordering. *Advanced Functional Materials* **18**(15), 2205–2214 (2008).
69. Xiong, S., Liu, Q., Wang, Q., Li, W., Tang, Y., Wang, X. & Chen, B. A flexible zinc tetrazolate framework exhibiting breathing behaviour on xenon adsorption and selective adsorption of xenon over other noble gases. *Journal of Materials Chemistry A* **3**(20), 10747–10752 (2015).

70. Banerjee, D., Zhang, Z., Plonka, A. M., Li, J. & Parise, J. B. A calcium coordination framework having permanent porosity and high CO_2/N_2 selectivity. *Crystal Growth & Design* **12**(5), 2162–2165 (2012).
71. Banerjee, D., Simon, C. M., Plonka, A. M., Motkuri, R. K., Liu, J., Chen, X., Smit, B., Parise, J. B., Haranczyk, M. & Thallapally, P. K. Metal-organic framework with optimally selective xenon adsorption and separation. *Nature Communications* **7**(11831) (2016).
72. Wu, T., Feng, X., Elsaidi, S. K., Thallapally, P. K. & Carreon, M. A. ZIF-8 membranes for Kr/Xe separation. *Industrial & Engineering Chemistry Research* **56**(6), 1682–1686 (2017).
73. Tozawa, T., *et al.* Porous organic cages. *Nature Materials* **8**(12), 973–978 (2009).
74. Jiang, S., Jones, J. T., Hasell, T., Blythe, C. E., Adams, D. J., Trewin, A. & Cooper, A. I. Porous organic molecular solids by dynamic covalent scrambling. *Nature Communications* **2**, 207 (2011).
75. Schneider, M. W., Oppel, I. M., Ott, H., Lechner, L. G., Hauswald, H. J. S., Stoll, R. & Mastalerz, M. Periphery-substituted [4+ 6] salicylbisimine cage compounds with exceptionally high surface areas: influence of the molecular structure on nitrogen sorption properties. *Chemistry — A European Journal* **18**(3), 836–847 (2012).
76. Mitra, T., Jelfs, K. E., Schmidtmann, M., Ahmed, A., Chong, S. Y., Adams, D. J. & Cooper, A. I. Molecular shape sorting using molecular organic cages. *Nature Chemistry* **5**(4), 276–281 (2013).
77. Chen, L., *et al.* Separation of rare gases and chiral molecules by selective binding in porous organic cages. *Nature Materials* **13**(10), 954–960 (2014).
78. Zhang, C., Lively, R. P., Zhang, K., Johnson, J. R., Karvan, O. & Koros, W. J. Unexpected molecular sieving properties of zeolitic imidazolate framework-8. *The Journal of Physical Chemistry Letters* **3**(16), 2130–2134 (2012).
79. Kwon, H. T. & Jeong, H. K. In situ synthesis of thin zeolitic–imidazolate framework ZIF-8 membranes exhibiting exceptionally high propylene/propane separation. *Journal of the American Chemical Society* **135**(29), 10763–10768 (2013).
80. Liu, J., Thallapally, P. K. & Strachan, D. Metal-organic frameworks for removal of Xe and Kr from nuclear fuel reprocessing plants. *Langmuir* **28**(31), 11584–11589 (2012).
81. Wang, H., Yao, K., Zhang, Z., Jagiello, J., Gong, Q., Han, Y. & Li, J. The first example of commensurate adsorption of atomic gas in a MOF and effective separation of xenon from other noble gases. *Chemical Science* **5**(2), 620–624 (2014).
82. Liu, J., Strachan, D. M. & Thallapally, P. K. Enhanced noble gas adsorption in Ag@MOF-74Ni. *Chemical Communications* **50**(4), 466-468 (2014).
83. Song, Q., Jiang, S., Hasell, T., Liu, M., Sun, S., Cheetham, A. K., Sivaniah, E. & Cooper, A. I. (2016). Porous organic cage thin films and molecular-sieving membranes. *Advanced Materials.*
84. Ryan, P. J., Farha, O. K., Broadbelt, L. J., Snurr, R. Q. & Bae, Y. S. Metal-organic frameworks for Xe/Kr separation (No. 8,784,536). Northwestern University, Evanston, IL (United States) (2014).
85. Maroulis, G. & Thakkar, A. J. Quadrupole polarizabilities and hyperpolarizabilities of Kr and Xe from fourth-order many-body perturbation theory calculations. *The Journal of Chemical Physics* **89**(12), 7320–7323 (1988).

86. Magdysyuk, O. V., Adams, F., Liermann, H. P., Spanopoulos, I., Trikalitis, P. N., Hirscher, M. & Dinnebier, R. E. Understanding the adsorption mechanism of noble gases Kr and Xe in CPO-27-Ni, CPO-27-Mg, and ZIF-8. *Physical Chemistry Chemical Physics* **16**(43), 23908–23914 (2014).
87. Wu, T., Lucero, J., Zong, Z., Elsaidi, S. K., Thallapally, P. K. & Carreon, M. A. Microporous crystalline membranes for Kr/Xe Separation: Comparison between AlPO-18, SAPO-34, and ZIF-8, *ACS Applied Nano Materials*, **1**(1), 463–470 (2018).
88. Anderson, R., Schweitzer, B., Wu, T., Carreon, M. A. & Gomez-Gualdron, D. Molecular simulation insights on Xe/Kr separation in a set of nanoporous crystalline membranes. *ACS Applied Materials & Interfaces* **10**, 582–592 (2018).
89. Sigma Aldrich — Xenon https://www.sigmaaldrich.com/catalog/product/sial/00472?lang=en&region=US.
90. Compressed Gas Association, *Handbook of Compressed Gases*, Reinhold Publishing Corporation: New York, NY (1966).
91. Darde, R. Prabhakar, Tranier, J. P. & Perrin, N. *Enrgy. Proced.* **1**, 527 (2009).
92. Betzendahl, R. *Cryo Gas International* **29**, 32 (2013).
93. Golden, T. C., Farris, T. S., Chiang, R. L., Whitley, R. D. & Taylor, F. W. U.S. Patent No. 6,658,894. 9 (2003).
94. Bazan, R. E., Bastos–Neto, M., Moeller, A., Dreisbach, F. & Staudt, R. *Adsorption* **17**, 371 (2011).
95. Parkes, M. V., Staiger, C. L., Perry, IV, J. J., Allendorf, M. D. & Greathouse, J. A. *J. Phys. Chem. Phys.* **15**, 9093 (2013).
96. Thallapally, P. K., Grate, J. W. & Motkuri, R. K. *Chem. Commun.* **48**, 347 (2012).
97. Wu, T., Lucero, J., Sinnwell, M., Thallapally, P. K. & Carreon, M. A. Recovery of xenon from air over ZIF-8 membranes. *Chem. Commun.* **54**(65), 8976–8979 (2018).

Chapter 9

MOF-based Mixed Matrix Membranes for Gas Separations

9.1. Introduction

The major concern with the polymeric membranes is its upper bound limitation as defined by Robeson, which was later confirmed by Freeman by providing the empirical correlation.[1–3] The main advantages of the polymeric membranes compared to other type of membranes are the processability and less cost of the polymeric materials. As mentioned in earlier chapters, porous inorganic membranes such as zeolite and metal organic framework membranes showed high potential for gas separation applications compared to the polymeric membranes.[4–12] Apart from high gas transport properties, the inorganic membranes also have very good mechanical, chemical and thermal properties. But these inorganic materials are very difficult and expensive process into different membrane configurations and cost of the materials is very high and therefore there are very few commercial inorganic membranes. Therefore, several research groups are working towards improving the processing conditions in different configurations using cheaper starting metal salts, organic salt and using water as the solvent. Meanwhile research was also focused on the alternative usage of these MOFs for gas separation applications. One of the most important and highly focused are the mixed matrix membranes (MMM) by embed them into the easily processable polymers.

MMM have been recently introduced as a new alternative approach that combines the benefits of both polymeric and inorganic materials. Initial studies of MMMs research focused on zeolites and carbon materials, but MMM research gained exponential growth after the invention of MOFs. A typical mixed-matrix membrane consists of fillers dispersed at a nanometer scale in a polymer matrix. The incorporation of inorganic fillers in the polymer matrix allows these membranes to inherit the superior separation performance from this material whereas the presence of a polymeric phase solves the inherent brittleness related problems to these fillers. In hybrid membranes such as MMMs, inorganic fillers act to create preferential permeation pathways that allow selective permeability imposing a barrier to undesired infiltrates in order to improve the selectivity towards a gas.

There have been several key issues related to fabrication of MMMs that have prevented the commercialization of these membranes. The fabrication of mixed matrix membranes presented several practical challenges at the polymer-filler interface due to the incompatibility of organic–inorganic interfaces.[13] There are several factors that are important in the design of a mixed matrix membrane. The compatibility of the polymer and filler particles in terms of their gas transport properties is very important and this will be discussed in detail in subsequent sections. Substantial improvement relative to the polymer occurs only when the permeabilities of the faster permeating component through the polymer and filler are approximately matched. If one phase has high gas transport resistance, it dictates the overall gas separation performance of the mixed matrix membrane. The other factor that influences the selection of the materials is the processing capability of the polymers after the addition of filler particles. Particle distribution within the matrix is another important factor to obtain the defect free membranes. The filler particles were also modified, especially in MOF based MMM, in order to improve polymer-filler interface and increased the gas transport properties. As stated above, initial evaluation of the component selection for the mixed matrix membrane can be done using the Maxwell equation with careful matching of the transport properties of each phase. Several other modified Maxwell model were develop to predict the gas separation performance of mixed matrix membranes.[14] For the mixed-matrix approach to achieve its full potential, the polymer matrix should have properties that lies close to the upper bound on the Robeson plot.

As seen from Figure 9.1, mixed matrix membranes (MMMs) are capable of overcoming the Robeson's Upper Bound through providing the selective gas transport pathways compared to polymeric and inorganic membranes.[12,15,16] Porous fillers act as molecular sieving agents separating gas

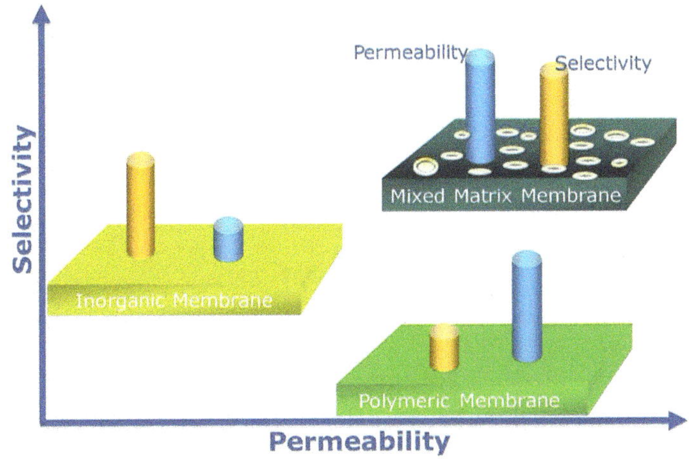

Figure 9.1. Comparison of gas transport properties of polymer membranes, inorganic membranes and mixed matrix membranes. (Reprinted from Ref. 15 with permission from Elsevier.)

molecules by their sizes or shapes.[10,12,17,18] These porous inorganic particles, such as MOFs, exhibit high gas fluxes and selectivities due to their small and precise apertures. These highly selective porous fillers allow passing the desired components making the permselectivity higher compared with a polymeric membrane. Thus, the addition of porous fillers to a polymer matrix results in an efficient way to achieve higher gas separation performance while making the membrane processing feasible. It not only solves the trade-off between selectivity and permeability but improves the thermal, chemical and mechanical stability.[19–21]

9.2. Fabrication and characterization of MMM

Mixed matrix membranes could have two types of morphology: symmetric and asymmetric, as shown in Figure 9.2. Symmetric dense membranes have been widely studied due to their simple synthesis process. In a typical synthesis process, the polymer solutions casted onto a glass plate using a knife and evaporate the solvent in the membranes under solvent saturated environment to form the dense membranes. Another disadvantage of this type of membranes is the thickness required for mechanical stability ($>50\,\mu$m) which reduces the gas flux through the membrane.

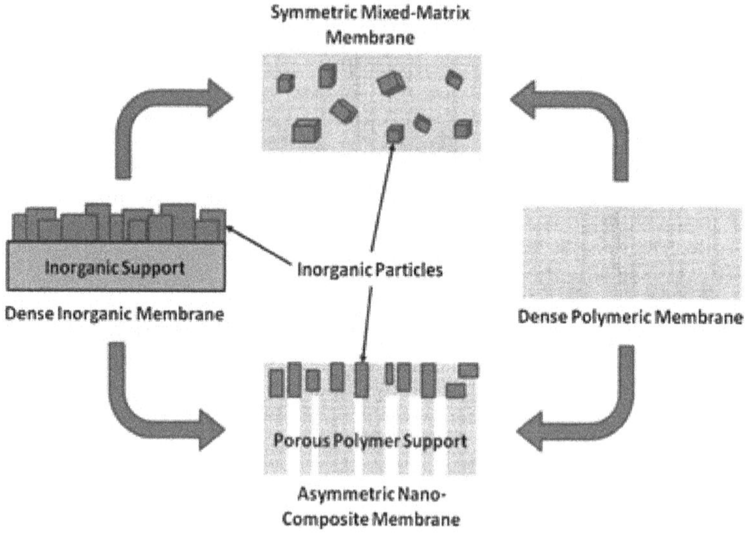

Figure 9.2. Schematic representation of different mixed-matrix membrane morphologies. (Reproduced from Ref. 21 with permission from the Royal Society of Chemistry.)

Contrary to symmetric membranes, asymmetric membranes have a thin ($<1\,\mu$m) selective skin used for gas separation and a nonselective porous support for mechanical stability. Asymmetric membranes are very thin compared to the symmetric dense membranes and the selective skin contains the dispersed inorganic filler.[19,21] Asymmetric membranes could be fabricated as flat sheet and hollow fiber configurations.[19] The asymmetric membranes can be synthesized by using several methods such as phase separation, interfacial polymerization, solution coating, plasma polymerization, etc.[22] As mentioned earlier first fabrication of asymmetric membrane with ultra-thin dense selective layer on a porous support was reported by Loeb and Sourirajan.[23] All industrial applications prefer asymmetric thin skin membranes because of their high gas fluxes while showing the intrinsic selectivity. Asymmetric hollow fibers are the most industrially practical geometry for gas separation membranes in terms of productivity and membrane packing efficiency. Although mixed matrix membranes show potential to be novel materials for high performance gas separation applications, several challenges have to be overcome in the design and fabrication of these membranes for industrial applications. The following sections discuss the different factors that effect the fabrication and performance of mixed matrix membranes.

9.2.1. *Polymer-filler compatibility*

In an ideal case, theoretically, MMM shows improved gas transport properties, in terms of permeability and selectivity compared to its polymer matrix. But in the real world, the dispersed phase may affect the surrounding polymer matrix at the interface between the two phases. Moore and Koros summarized several different non-idealities that arises during the formation of MMM as shown in Figure 9.3. As shown in Figure 9.3, the interfacial defects can be categorized into five cases as presented by Moore and Koros:[13] (1) rigidified polymer layers wrapped around the fillers, (2) large voids (>~10–100 nm) at the polymer/filler interface, which could be an unfavorable result after the MMM processing, (3) small voids at the interface mainly because of the rigidity of polymer chains and less attractive interaction between polymers and fillers, (4) clogged fillers by gas molecules

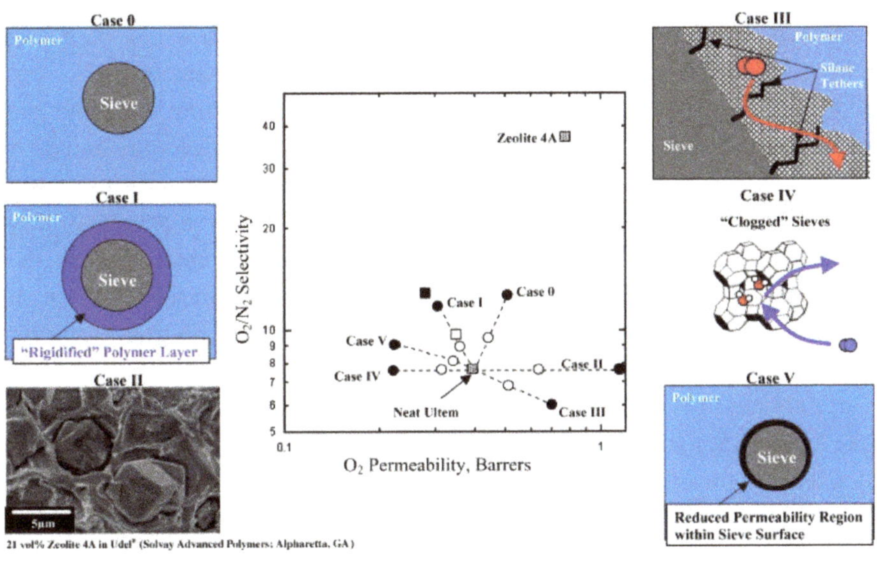

Figure 9.3. Summary of the relationship between mixed matrix membrane morphologies and transport properties. Circles represent calculated values; squares represent experimental in Ultem data. Solid markers are 35 vol% zeolite 4A; open markers are 15 vol% 4A. (Reprinted from Ref. 13 with permission from Elsevier.)

because of strong holding from the inside surface of fillers, and (5) bottle-necked at filler side of the interfacial region.

In order to avoid these non-ideal behavior, the polymer and filler particle needs to be compatible both in terms of chemical properties and gas transport properties. One way to achieve ideal interface is by choosing polymer and inorganic materials with the similar polarity.[24] The interaction forces between the polymer and filler should be higher than both polymer-solvent interaction and filler-solvent interaction in order to achieve good interface. The additional requirement of polymer flexibility during formation would be harder to achieve because polymers close to the upper bound are inherently rigid, but this can still be achieved by the use of suitable plasticizing mechanisms. In particular, it is desirable to prepare membranes in such a manner that the stress at the organic–inorganic interface is minimized to suppress the tendency to form undesirable voids at that interface. Other methods include the casting of the membranes at a temperature above or at glass transition temperature,[25,26] annealing of membranes above glass transition temperature,[27] adding surface coupling agents,[28,29] preparation of membranes using melt processing,[13,30] using soft-hard block copolymer,[31] etc. Furthermore, processing steps and solvents that can 'clog' the zeolites should be avoided.[32] As mentioned in the introduction, it is very important to match not only chemical properties of polymer and filler but also the gas transport properties in order to achieve improved gas transport properties. Moore and Koros made a great conclusion that the substantial improvement in MMM relative to the neat polymer can be achieved only when the permeabilities of the faster permeating component through the polymer and sieve are approximately matched.[13] The careful selection of the polymer and the filler materials greatly impacts the resulting morphology and performance of the MMM. The appropriate choice of both of these materials requires a clear understanding of the transport mechanisms. Pore size and surface functionality of the filler particles are also very important in selecting the filler materials to achieve high gas transport properties in MMM. Most of the fillers that used in the formation of MMM are microporous materials such as zeolites, carbon molecular sieves and metal organic frameworks. Other fillers that were used in the mixed matrix membranes as fillers are silica, activated carbon, metal nanoparticles, graphene, carbon nanotubes, porous polymers, covalent organic frameworks, etc.[19–21,32–40]

The microporous materials shows very high selectivity owing to their molecular sieving capability compared to the nonporous fillers. Another factor that effects the gas transport properties of the MMMs are the surface functionality of the filler particles which improves adsorption of the gases on their surfaces which in turn results in improved molecular diffusion through its pores. Rigid materials with pore sizes bigger than the penetrants have also been studied as well. In this case, the most adsorption capacity or condensable component will adsorb and diffuse more effectively through the particles while the less adsorption capacity will permeate slowly.[41] When using these large pore size rigid materials, it is the gas feed mixture that determines surface flow or molecular sieving as the dominant mechanism for transport. The nonporous fillers improve the gas separation performance of polymers by increasing the matrix tortuous pattern and decreasing the diffusion of larger molecules.[37]

9.2.2. *Morphological properties of the MOFs*

The typical membrane thickness for asymmetric membranes should be much less than 1 μm in order to be applicable in the industry. The thickness of the membrane layer in mixed matrix membranes largely depends on the particle size. In order to achieve very thin selective layer, the particles must be much less than the layer thickness. Also, the smaller particles have higher surface area/volume ratios that normally equate to enhanced mass transfer between the phases. Nano-scale MOFs may alter the packing of the polymer chain increasing the free volume between chains resulting in an increase of the gas diffusion which results in an increase in the permeability of the penetrants.[34] The effect of particle size on the gas separation properties of MOF based MMMs was studied by several groups. There is one study where they systematically changed the particle size of silicalite in Polydimethylsiloxane (PDMS). They found that the permeability decreased with a decrease in particle size which was contradictory to the theoretical prediction. They hypothesized that this result was due to enhanced solid/polymer interfacial contact or densification of the polymer around the particle because of the rubbery nature of PDMS. Su *et al.* also studied the particle size effect by using different size silica nanoparticles and dispersing them in cross-linked polyethylene oxide polymer.[42] They show that the gas separation performance of MMM deviates from the theoretical values as the particle size decreases.[43] We need to take utmost care while using smaller MOF particles to keep them segregated and dispersed well within the polymer matrix.

A critical characteristic in the fabrication of MMM is the limit of particle loading which should not exceed 50% due to agglomeration issues resulting in the gas transport dominated by the polymer phase.[21,44] In most cases, the higher loading results in lower selectivity due to the defect formations around the agglomerates. The volume of particle loading also depends on the particle size. As the particle size goes down, the weight percent that can be added into the polymer decreases. In most cases, the increase in particle loading above the critical weight percentage results in the agglomeration of particles and reduction in the mechanical properties.[27] It has been observed that at low loading of particles, the viscosity of the solution is low and hence the dispersion of the particles in the polymer phase is not uniform.[36] While at high loadings increase the viscosity, sedimentation occurs resulting in defects, then the loading must be optimized in order to have appropriate MMM. Agglomeration is very common at high temperatures which could cause wide gaps and voids between the filler and polymer. This could also occur due to the surface tension and buoyancy of the processing fluid.[36]

In the case of MMM, the polymer-filler solution needs to be homogenous in order to well-disperse the inorganic filler in the membrane structure. This can be achieved by dispersing the inorganic particles in the solvent using ultrasonication horn before adding the polymer. Koros *et al.* developed a technique called *Priming* in order the keep the filler particles dispersed in the polymer solution.[45] It was observed that particles should be added into the solvent in order to disperse them well using the mixer. A small amount of polymer addition to the particle dispersed solution avoids the agglomeration of particles by wrapping them with the polymer chains. Large particle size results in the sedimentation or setting of particle at the bottom of the film. This can be avoided by using the nano-size filler particle which stay in suspension for longer periods. The settling can also be reduced by increasing the viscosity of the solution. A recent review by Dong *et al.* clearly explained several other methods used for the dispersion of particles in the mixed matrix membranes.[21]

Most of the research reports were focused on improving the interface between the particle and polymer. These reports used several techniques as mentioned above such as surface modification of filler, using of ionic liquids, melt processing, using of plasticizers, etc. Most of these techniques showed good improvement compared to the MMM that were made without any modification.

9.2.3. *Improving the polymer-MOF surface interaction*

The filler and polymer matrix compatibility and the interaction between polymer-sieve-solvent are the major intrinsic factors affecting the structure of the MMMs.[46] In earlier MMM research using zeolites, surface modification of particles using coupling agents have been extensively used especially in the case of zeolites.[19,28,36,38,47–51] One end of the silane agent can be reacted to the hydroxyl groups on the surface of zeolites while the other end attaches to the polymeric chain. In this way, the silanes act as a bonding agent between the polymer and the particles improving the interface between the polymer and zeolite. General procedures and mechanisms for silanization of a zeolite have been described elsewhere.[47,49] Furthermore, the addition of hydroxyl and other functional groups on the surface of these nanomaterials may lead to interactions with polar gases such as CO_2 improving the penetrant solubility in the MMMs.[52] The simplest approach to enhance the link between the polymer and the inorganic particles is the introduction of a third component. For this purpose, low molecular additives are used such as 2,4,6-triaminopyrimide (TAP) and p-nitroaniline (PNA).[53,54] This approach leaded to a better interface and higher selectivity and permeability comparing to unmodified particles. It is clear from the SEM images that the contact between the zeolite and the polymer is better for the modified zeolites.

Due to the incompatibility between the polymer and inorganic phases that can lead to non-ideal morphologies of the synthetized mixed matrix membranes, there are several casting conditions that have been extensively followed. One of these widely used approaches is to perform the casting at temperature above the polymer glass transition or either use a polymer with low glass transition temperature and perform the membrane fabrication close to the polymer glass transition temperature.[27,47,49,55,56] To maximize stress relaxation in this method the matrix flexibility is maintained during the membrane formation. Thus, a low volatile solvent is the choice. However, such a solvent is difficult to find since the typical solvents used for film casting boil below this temperature.[13,27] An alternative is annealing above the glass transition temperature. However, the defective membranes formed using this method do not lead to significant improvements.[27] To maintain polymer chain flexibility and mobility and reduce the glass transition temperature during the membrane synthesis it is common to incorporate a plasticizer into the polymer solution, such as RDP Fyroflex, 4-hydroxy benzophenone and di-butyl phthalate.[27,53]

Although metal-organic frameworks are intrinsically more compatible with polymer because of their organic nature of the structure, several researchers modified them with different functional groups to tune its compatibility with polymer. UiO-66 MOF was modified with several organic functional groups such as amione, phenylacetyl chloride, decanoyl chloride, succinic anhydride and improved the interface properties.[16,57,58] Work to improve the MOF/polymer interface was continued in 2015 by Venna, *et al.* when they modified the surface of (Zr) UiO-66-NH_2 MOF particles to improve its interaction with the polymer matrix, Matrimid®.[57] The MOF is composed of $Zr_6O_4(OH)_4$ clusters bridged with 2-amino-1,4-benzenedicarboxylates that form triangularly shaped pore apertures of 0.6 nm. (Zr) UiO-66-NH_2 was used due to its water, chemical, and thermal stability, and its ability to undergo post-synthesis chemical modification without degradation. The authors rationalized that the MOF/polymer interface could be improved by analyzing the structure of the polymer to determine the functional groups that could form an ideal interface with the MOF and then chemically modify the surface of the MOF with such functional groups. The authors prepared four MOFs (Figure 9.4) having polar, non-polar, or aromatic surface functional groups: neat (Zr) UiO-66-NH_2 (I), aromatic-modified (Zr) UiO-66-NH_2 (IPA), aliphatic C_{10}-modified (Zr) UiO-66-NH_2 (IC10), and acid-modified (Zr) UiO-66-NH_2 (ISA). The synthesized MOFs were activated with Soxhlet extraction for 7 days with Cl_2CH_2.

The degree of functionalization of the MOFs was calculated from LCMS, XPS, and 1H NMR data: the percentages of functionalized ligands obtained were 2%–6% for IPA, 2%–5% for IC10, and 16%–32% for ISA. Owing to the size of the phenyl and alkyl chains that limit their diffusion into the channels of the MOF, the authors expected the surface functionalization for IPA and IC10 to be low. The higher level of ISA functionalities, however, indicated that some internal functionalization occurred along with

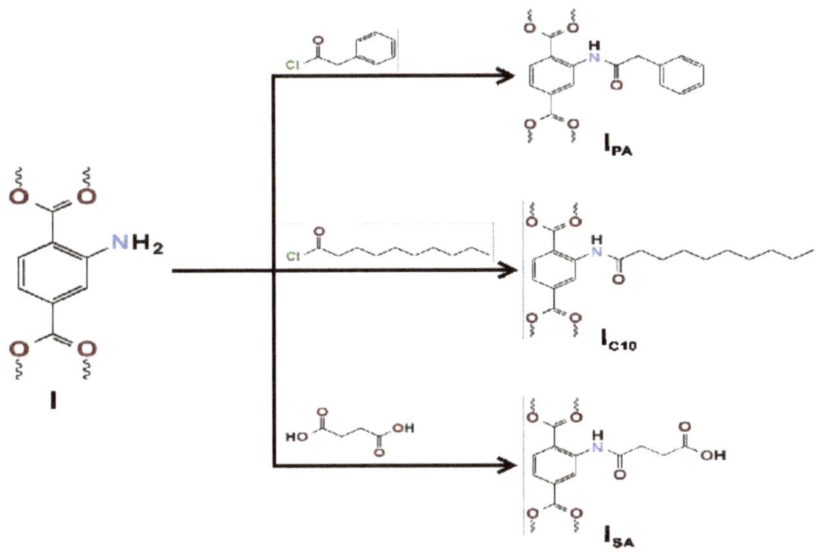

Figure 9.4. Post-synthesis modification of (Zr) UiO-66-NH2 (I) for functionalization at the –NH2 functional group of the linker with aromatic (IPA), aliphatic C10 (IC10), and acid (ISA) terminal functional groups. (Reproduced from Ref. 57 with permission from the Royal Society of Chemistry.)

surface functionalization. MMMs of Matrimid® and (Zr) UiO-66-NH_2 (and its functionalized derivatives) with loadings of 12 to 40 wt% were prepared with an initial priming of the MOF with a polymer solution. The MMMs were cast on glass, dried at 100°C, and then annealed at 225°C for 2 h in a vacuum oven.

Pure CO_2 and N_2 gas permeation studies were performed at room temperature and 1.38 bar. The permeabilities and CO_2/N_2 selectivities of the polymer and the 23 wt% (Zr) UiO-66-NH_2 (I, IPA, IC10, ISA)/Matrimid® MMMs are shown in Figure 9.5. The results indicated that CO_2 permeability increased largely for all MMMs with respect to the polymer, but only the MMMs containing the neat and the IPA-functionalized MOF exhibited increased CO_2/N_2 selectivity relative to the polymer, which was attributed to the formation of a defect-free interface between the MOF and the polymer and the improved surface diffusion of CO_2, since it adsorbed more favorably to the MOF than N_2. The authors also concluded that MMMs containing the IC10 and ISA functionalities did not increase CO_2/N_2 selectivities

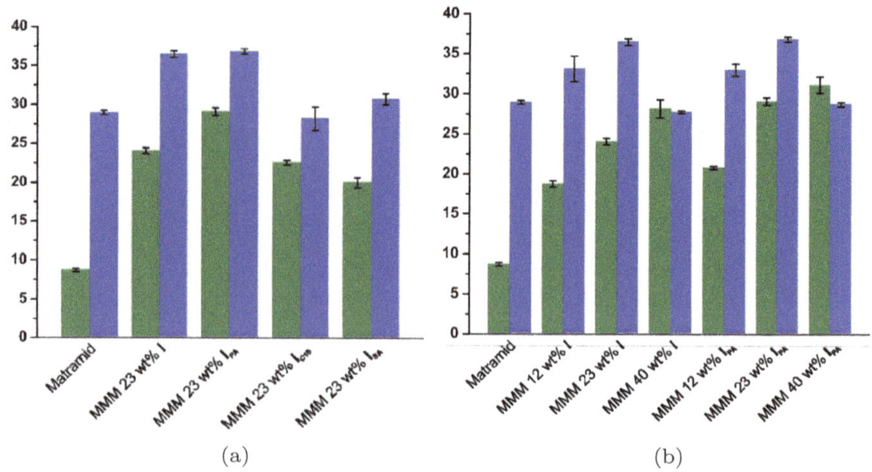

Figure 9.5. (a) CO_2 permeabilities (green) and CO_2/N_2 ideal selectivities for the polymer and the 23 wt% (Zr) UiO-66-NH_2 (I, IPA, IC10, ISA)/Matrimid® MMMs; (b) CO_2 permeabilities (green) and CO_2/N_2 ideal selectivities for the polymer and the 12, 23, and 40 wt% (Zr) UiO-66-NH_2 (I, IPA)/Matrimid® MMMs. (Reproduced from Ref. 57 with permission from the Royal Society of Chemistry.)

because of the defects created at the MOF/polymer interface due to incompatibilities between the polymer and the IC10 and ISA functional groups on the MOF. Additional testing with 12, 23, and 40 wt% (Zr) UiO-66-NH_2 (I, IPA)/Matrimid® MMMs revealed that the membranes increased the permeability for CO_2 with increasing MOF loadings (Figure 9.5b). CO_2/N_2 selectivities, however, reached a maximum at 23 wt% and then declined at 40 wt% signaling the presence of interfacial defects induced by the agglomeration of the MOF crystals at high loadings. In the case of IPA functionalized MOFs, the aromatic groups present in the Matrimid are expected to interact favorably through π-π stacking with the aromatic ring of the –PA group and the imide groups in the polymer are expected to interact strongly with the –NH_2 groups and amide linkages through hydrogen bonding as shown in Figure 9.6.

Ti-exchanged UiO-66 MOFs were used to increase the interface and performance of the mixed matrix membranes using PIM-1 as the polymer matrix.[59] MIL-based MOFs were also modified with several chemical surface groups and metal exchange methods which resulted in the improved interactions between the MOF and the polymer.[60–64]

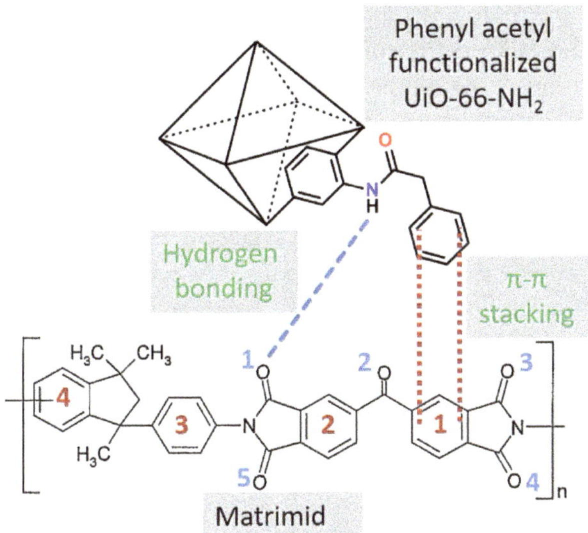

Figure 9.6. Scheme demonstrating the favorable interactions between the Matrimid® polymer and IPA based on surface functionality. (Reproduced from Ref. 57 with permission from the Royal Society of Chemistry.)

MMMs have also been prepared by combining a small amount of highly absorbing non-toxic ionic liquid and MOFs in order to avoid the void structure while improving the gas separation performance by utilizing the high CO_2 absorption of capacity of ionic liquids.[65] The ionic liquid, being charged and non-volatile, can remain within a polymer matrix prepared using polymerizable ionic liquids. Addition of sufficient ionic liquid to wet the zeolite keeps the particles dispersed within the matrix. An ionic liquid has been used to wet both the polymer phase and a zeolite (SAPO-34) for selective gas separations.[66–68] Another method is to perform the casting on a dense liquid surface with high surface tension in order to avoid strong membrane adhesion to the casting substrate surface.[13] Herein, the polymer solution can contract freely after casting and the stresses distribute uniformly in every direction and forms void-free membranes. Melt processing is another method used to maintain the polymer flexibility during membrane preparation; however it cannot lead to feasible industrial relevant membranes.[30] Furthermore, low molecular weight materials or bridging agents could be used to fill the voids between the inorganic filler and the polymer phase. These materials must be able to form hydrogen bond, hydroxyl or carbonyl bond besides being soluble in the solvent while making the polymer dope solution. Long aliphatic and polyatomic-based compounds with polar atoms, rigid and planar structure serves this purpose; p-nitroaniline (pNA), 2,4,6-triaminopyrimidine (TAP) and 2-hydroxy 5-methylaniline are a few examples.[36,38,69]

Another way is crosslinking of the copolymers through acetal linkages which could provide the flexible component of rubbery polymer and good contact for filler and rigidity of the glassy polymer. No additional substances must be added in this method.[36] The use of some copolymers to enhance the interfacial contacts polymer-particle has been also explored. This method allows obtaining void-free membranes. However, finding an industrial suitable copolymer or further synthetizes increases the number of preparation steps.[31]

The minimization of MOF-solvent/non-solvent interaction method is another way to obtain interfacial void-free MMMs. Hydrophobization of the filler surface by replacing the hydroxyl groups with methyl groups avoid the particles which behave as nucleating agents to improve the adhesion.[48] Alternatively, one can use a hydrophobic material for both phases, inorganic and organic. An example of this method is the zeolite Nu-6(2) and polysulfone. MMMs synthetized using these latest can result in free-void interfaces.[70] Another method called *sizing technique* which uses the sizing agents (typically polyvinylpyrrolidone) to increase the strength of the interface by increasing the chemical reactive site and the surface area for adhesion without using complicated chemical reactions.[71,72]

Marti *et al.* fabricated MMMs using in-situ MOF growth method to avoid the defect formation and minimize the MMM fabrication steps. Nano-UiO-66 was synthesized in Matrimid matrix in a single step MMM fabrication. All the precursors needed for UiO-66 synthesis were dissolved in Dimethylformamide (DMF) solvent and Matrimid polymer was also dissolved in the same solution. This solution was heated at around 120°C for a specific to form the UiO-66 MOF while drying off the solvent resulting in the formation of MMM as shown in Figure 9.7. Using this method, many synthetic steps that are necessary for traditional MMM fabrication can be avoided, saving time and reducing cost, which could be an enabling technology for large.

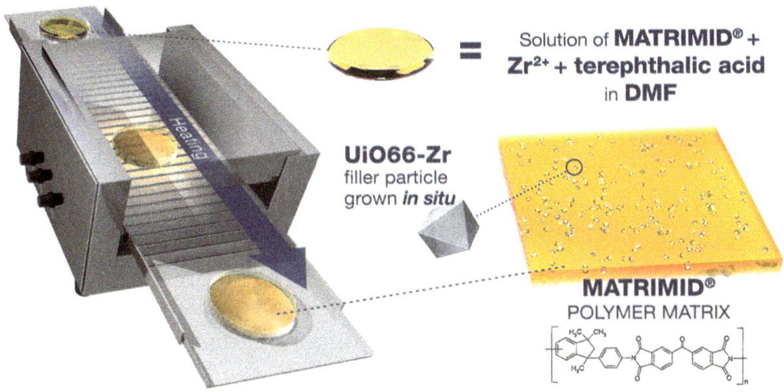

Figure 9.7. Fabrication of MMMs by growing UiO-66 MOF in-situ within the Matrimid® polymer matrix in a single step. (Reprinted from Ref. 73 with permission from the American Chemical Society.)

9.2.4. *Membrane module configurations: Flat sheet and hollow fiber*

MMM can be fabricated as flat sheet membranes and hollow fiber membranes. Initial studies on the MMM were solely focused on the fabrication of flat sheet membranes. Flat sheet membranes have been used widely in gas separation industry and have been exhaustively studied for the past 20 years despite the disadvantage of dense selective layer and lower gas permeation flux. For symmetric flat dense membranes the thickness would be more than 50 μm and for asymmetric, it could go from 5 to 25 μm.[19] Recently, flat sheet membranes have attracted increase attention for potential application in water desalination,[74,75] and waste water treatment.[76–78] Most of the flat sheet mixed matrix membranes were fabricated as dense membranes. There are very few research reports that focused on the fabrication of asymmetric flat sheet membranes.[79,80] Basu *et al.* prepared asymmetric Matrimid®/[$Cu_3(BTC)_2$] mixed-matrix membranes using phase inversion technique to improve the membrane gas separation performance.[79] The selectivity and permeance increased with increasing filler loading compared to pure polymer membranes.

Recently, researchers were focused on the fabrication of hollow fiber mixed matrix membranes because of their intrinsic advantages such as high surface area to volume ratio. Furthermore, high selectivity, good mechanical strength, good chemical resistance, low fouling, thermal stability, cost effective, defect free production and good compatibility makes them more flexible during operation and handling in the module fabrication.[25,36,76] Despite these advantages, there are several challenging factors related to the fabrication of hollow mixed matrix membranes such as the thickness of the membrane (should be as thin as possible), low defects in the selective thin skin, thin dense separating layer which lead to faster production (but must be able to hold the pressure during operation) and sub-micro sized particles which could fit inside the ultra-thin layer and suppressed polymer and polymer-particle defects, which are important features of Hollow fiber asymmetric MMMs. It has been demonstrated that in hollow fiber membrane, the particle distribution should be controlled by manipulating the air gap or draw ratio instead of high loading in order to avoid processing problems.[19]

Typically hollow fiber MMMs are produced by single-layer spinning where phase inversion method is used and the inorganic filler is introduced in the spinning polymer.[48] Figure 9.8 shows the typical structure of hollow

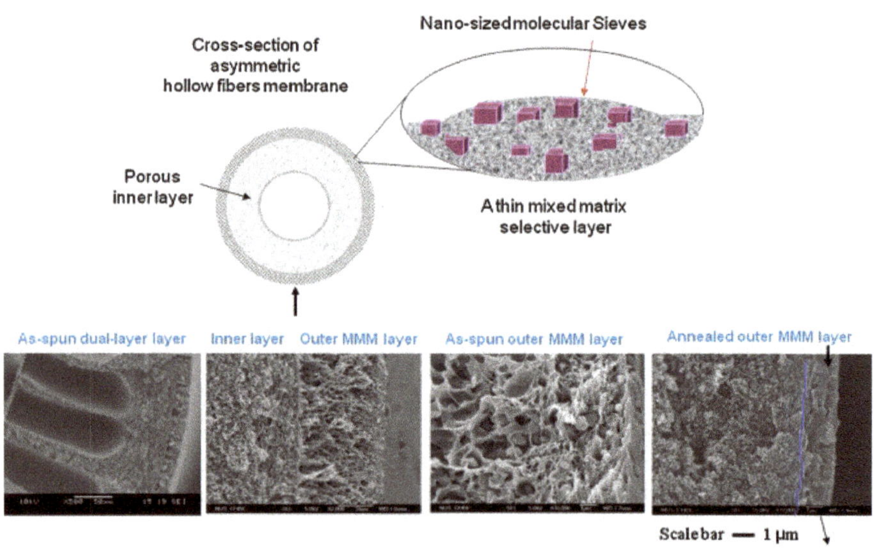

Figure 9.8. Schematic cross-section morphology of the dual-layer hollow fiber with a polymer–particle mixed matrix skin. (Reprinted from Ref. 81 with permission from Elsevier.)

fiber mixed matrix membranes with a thin skin layer in which inorganic filler particles were fully embedded in it and the porous support which provides the mechanical support to the thin MMM.[81] Another common method used is dual-layer spinning method. Dai *et al.* fabricated mixed matrix asymmetric dual layer hollow fiber membranes using dry jet-wet quench process containing the metal-organic framework sieve ZIF-8 in an Ultem® polymer matrix.[82] Dual-layer fibers were produced via co-extrusion of the core and sheath dopes with a bore fluid solution through a composite spinneret. There are several advantages of this method such as high performance using cheaper polymer as support, control over the porosity of the membrane, easy fabrication and higher fluxes using single-step production. This method also add more complexity in designing and controlling the spinning parameter. Also it is very important to note here that the high gas separation performance polymers which are non-spinnable can be used as the outer coating layer and achieve high performance. Also this method needs much lower amount of fillers compared to single layer hollow fiber membranes. Usage of small quantities of fillers is very important to produce larger area modules especially in the case costly MOFs.

9.3. Gas separation performance of MOF-based MMM

Porous solids with higher permeability and selectivity than polymer are highly desirable as additives in MMM. As mentioned above, many different type of particles such as zeolite, MOF, carbon molecular sieves, porous organic polymer (POP), covalent organic framework (COF) and nanoparticles were used to improve the gas separation performance of mixed matrix membranes. Polymer and porous particle interfacial interactions such as pore blockage, chain rigidification, free volume increase, and interfacial voids formation dictates the gas transport of MMM.[13]

The organic ligands as an inherent part of MOFs allow them to interact well with the polymers compared to the zeolites and avoid the non-ideal interfaces. More importantly, size, shape and chemical functionalities of the MOF cavities can be easily tailored by choosing the specific organic linker in the synthesis or by post-synthetic functionalization and thereby improve the filler-polymer interaction.[83] By selecting the proper MOF and polymer i.e. by matching the gas separation properties and filler-polymer interactions, it is very easy to surpass the Robeson's Upper Bound.[13] MOFs also have high pore volumes and low density than the zeolites and affect the gas separation performance of mixed matrix membranes at lower loadings compared to zeolites. Gas transport in the mixed matrix membranes is mainly governed by solution-diffusion mechanism since polymer volume fraction is much higher than the MOF particles.

The most important feature of MOF that makes them more attractive for MMM application compared to other filler particles is its intrinsic compatibility with the polymers because of its organic linkers and their functionalities. Another important feature is its porosity and surface area is much higher than the zeolites which makes them the most suitable additives for mixed matrix membranes. The availability of different MOF with outstanding properties leads to rapid growth of MOF in MMMs. Some of the examples of MOF-based MMM are shown in Table 9.1. First fabrication of MOF-based MMM incorporated copper (II) biphenyl dicarboxylate–triethylenediamine MOF in the poly(3-acetoxyethylthiophene matrix for gas separation.[90] The permeability of CH_4 is enhanced with MOF loading. The authors claimed that the increase adsorption of methane in the copper MOF is responsible for increase in CH_4 permeability.[90] Several MOF types based on Cu, Zn, Mn, Al, Cr metal ion and benzene-1,3,4-tricarboxylate (BTC), benzene-1,4-dicarboxylate (BDC), 2-methyl-imidazolate, 2-carboxyaldehydeimidazolate, 2-amino-terephthalate linkers are developed.[20] Perez *et al.* summarized the usage of several MOFs composed of different organic linkers and metal MOFs for the fabrication of MOF-based MMM.[91]

Cu-BDC MOF synthesized from copper nitrate trihydrate and terephthalic acid is added to polyvinyl acetate (PVAc) matrix. 15 wt% Cu-BDC-MOF loading PVAc/Cu-BDC MOF MMM showed improvements in the pure gas permeabilities of He, CO_2, O_2, N_2 and CH_4 along with enhancements in He/CH_4, O_2/N_2, N_2/CH_4, CO_2/N_2 and CO_2/CH_4 selectivity. The permeation time lag behavior of the MMMs suggested good PVAc-Cu-BDC MOF adhesion as well as easy penetrant access to the channels of Cu-BDC.[36,84] Car *et al.* incorporated $Cu_3(BTC)_2$ (copper(II)–benzene-1,3,5-tricarboxylate) and $Mn(HCOO)_2$ (manganese (II) formate) in polysulfone (PSf) polymer matrix.[92] MMM containing MOF showed more or less higher permeabilities than the pure PSf, while improvement in selectivities as a consequence of increase diffusivity factor.[92]

Table 9.1. Performance of some of the polymer-MOF mixed matrix membranes.

	Pure polymer					Mixed matrix membranes					
	Permeability (Barrer)			Selectivity		Permeability (Barrer)			Selectivity		
MMMs	N_2	CO_2	CH_4	CO_2/N_2	CO_2/CH_4	N_2	CO_2	CH_4	CO_2/N_2	CO_2/CH_4	Ref.
Polyvinylacetate/ Cu-BDC (15 w/w)	0.078	2.44	0.068	31	35.0	0.09	3.26	0.80	35.7	40.4	84
Matrimid/ZIF-8 (60% w/w)	0.31	9.52	0.24	30.6	39.8	0.44	8.08	0.10	18.3	80.7	44
SPEEK/MIL-101(cr) (40% w/w)	0.48	15.8	0.69	33.0	27.2	0.83	31	1.0	37.0	31.0	85
Poly(trifluoroethoxy) Polyphosphazene/ 20% SIFSIX-Cu-2i	24.6	325	—	13.2	—	23.5	441	—	18.8	—	86
Matrimid/ NH_2-MIL 53(Al) (8% w/w)	—	8.01	0.20	—	37.3	—	9.03	0.25	—	36.5	86
Matrimid/NH_2-MIL-125 (30% w/w)	—	6.00	0.20		30.0	—	50.0	1.35	—	37.0	87
Matrimind/UiO-66-NH_2 (23% w/w)	—	8.5	0.29	29	—	—	29.0	0.78	37.0	—	57
Polysulfone/NH_2-MIL-125(Ti) (30 wt%)	—	9.5	0.43	—	22	—	40	1.36	—	29.2	88
Matrimid/MIL-53 (37% w/w)	0.25	8.4	0.22	33.6	38.1	1.8	51.0	1.1	28.3	46.4	89
6FDA-MMM	28	390	16.2	14	24	32.7	720	19.5	22	37	99
Polymers of Intrinsic Microporosity-1/5% Tix UiO-66	180	3620	240	20	15	660	13540	1220	20.5	11	59

Zeolite imidazolate frameworks (ZIFs) are another MOF type synthesized from the linker containing imidazolate and Zinc (Zn), named as ZIF due to their resemblance to zeolite topologies and the metal-imidazolate-metal bond angles to the zeolite Si–O–Si angles. ZIF-8 (Zn(2-methylimidazole)2) prepared from Zn metal and 2-methylimidazole linker is most widely studied MOF in MMMs. ZIF-8 embedded into PSf matrix showed improved gas adsorption in the MMM with increase in ZIF-8 loading.[93] ZIF-8/Matrimid® MMMs are fabricated with loading up to 80% (w/w) and observed loss of mechanical strength and decrease in flexibility.[44] The permeabilities of H_2, CO_2, O_2, N_2, CH_4, C_3H_8 are increased with increase in ZIF-8 loading up to 40% (w/w) and decreased for higher loading of 50% and 60%(w/w). This result indicates a transition from a polymer-driven to a ZIF-8-controlled gas transport process, where at higher loadings than 40% (w/w) evidenced the sieving effect. The ideal selectivities for a gas pair containing small molecules, such as H_2/CH_4, CO_2/CH_4, H_2/C_3H_8, and CO_2/C_3H_8 are increased consistently with increase ZIF-8 loading, suggesting molecular sieving.[44]

MIL-53 MOF type synthesized from Al^{3+} or Cr^{3+} metal ions and benzene-1,4-dicarboxylate (BDC) linker is a special type of flexible MOF than can reversibly contract or open the pores upon the adsorption of molecules like CO_2 and H_2O.[94] Basu *et al.* incorporated MIL-53 (Al) MOF in Matrimid® and observed 218% increase in CO_2 permeance compared to pure Matrimid® membranes at 30% (w/w) loading. Selectivity of CO_2/CH_4 increased up to 15% at 30% (w/w) loading indicating good interaction between MIL-53(Al) and Matrimid®.[95] NH_2-MIL-53(Al) MOF synthesized from Al^{3+} ion and 2-amino-terephthalate are embedded in polysulfone and exhibited moderate increase CO_2 permeability up to 40% (w/w) loading and CO_2/CH_4 selectivity increased two fold up to 25% (w/w). The significant enhancement in the selectivity can be related to good interaction between amino functionalized framework and polysulfone due to hydrogen bonding between amine and sulfone groups at their interface.[96]

Thompson *et al.* prepared mixed matrix membranes containing mixed organic linker ZIF materials with ZIF-8 crystal topologies in order to increase the selectivity.[97] This work has shown that tuning of the organic linker composition in mixed-linker ZIF materials allows control of permeation of gases in mixed matrix membranes and ultimately resulted in improved gas separation properties with CO_2/CH_4 selectivities up to 40. Mixed matrix membranes were prepared using various MOFs such as ZIF-90, NH2-MIL-53, ZIF-8, ZIF-7, IRMOF-5, Cu_3BTC_2, etc.[98] Most of the MOF- based mixed matrix membranes were prepared using glassy polymers like polyimide. Some MMM were also prepared using the rubbery polymers like PDMS. Flexibility of the chains in the rubbery polymers resulted in good interaction with the MOFs and showed that gas separation performance increases with the incorporation of MOFs.[92] But the increase in gas separation was not very high because of the comparable permeability of the rubbery polymers and the filler particles. But in the case of glassy polymers, the improvement in the performance is very pronounced because of the incorporation of highly permeable filler particle in the low permeable glassy polymers. Formation of defects in the mixed matrix membranes was the main problem in using the glassy polymers because of the rigid polymer chains. The separation performances of different MOF and polymer combinations are given in Table 9.1.

Bae *et al.* prepared the mixed matrix membranes by dispersing ZIF-90 fillers in the low permeable polymer like Matrimid and high permeable polymer like 6 FDA-DAM polymers and studied the gas separation performance.[99] There was only a slight increase of permeability with the mixed matrix membranes containing Matrimid (~20% increase) but in the case of 6 FDA-DAM, the permeability was almost doubled as shown in Table 9.1 and showed the gas separation performance above the Robeson's Upper Bound. So, it is important to match the gas separation properties of polymer and the MOF in order to achieve the optimized gas separation performance using the mixed matrix membranes.[13] Nafisi *et al.* prepared the ZIF-8-based MMM with the PEBAX polymer and also showed very high CO_2 permeabilities up to 1,287 while retaining the selectivity.[100] Compared to the other literature reports on mixed matrix membranes, these membranes showed highest gas separation performances. Perez *et al.* showed that MOF-5/Matrimid® mixed matrix membranes showed no significant change in solubility with increased MOF-5 loading compared to pure Matrimid®, but the permeabilities of the gases tested increased 120% while the ideal selectivities remained constant at 30% MOF-5 loading in the Matrimid®.[101] Therefore, the permeability of the gases is enhanced by their increase in diffusivity in the membrane owing to the porosity of the MOF-5. These results are consistent with

Freeman's conclusions that diffusivity plays a major role compared to solubility in achieving higher permeability.[3] Yang *et al.* showed increased H_2 permeability and H_2/CO_2 selectivity with the incorporation of ZIF-7 in the PBI polymer and attributed this increase to the introduction of free volume due to the strong interfacial interactions between the filler and the particle which accelerated the gas transport and the rigidified chains enhanced the H_2 gas selectivity.[102] This work also prevented the agglomeration of particles and successfully minimized the nonselective voids in the membrane which resulted in the higher gas separation performance.

Much research also focused on the functionalization of the MOF particles in order to increase the interaction and improve the separation performances. Rodenas *et al.* prepared the MMM using the amine functionalized flexible MOF filler MIL-53. These mechanically-improved membranes showed promising CO_2/CH_4 gas separation performances as high as 70% enhancement in CO_2 permeability combined with a slightly better separation factor compared to neat polymer counterparts.[103] Zornoza *et al.* also prepared the MMM by dispersing the amine functionalized MIL-53 in polysulfone and showed increased performance even at high pressure. This work also observed that CO_2/CH_4 selectivity increases with pressure, in contrast to many reports, due to the intrinsic flexibility of the filler particles or the breathing effect.[104] Amine functionalized MOF-199 and UiO-66 also showed improved interaction and gas separation performances. Due to the presence of $-NH_2$ functional groups in the MOF structure lead to creating rigidified polymer at the interface of the filler and polymer matrix which resulted in decreased permeability and increased selectivity.[16] Another important development in the field of MOF-based MMM is the preparation of asymmetric mixed matrix membranes (MMMs). Basu *et al.* prepared the defect-free asymmetric MMM via phase inversion using the 1:3 blend of Matrimid®/Polysulphone and $Cu_3(BTC)_2$ MOF with improved thermal and mechanical properties. The CO_2/CH_4, CO_2/N_2 selectivity and CO_2 permeance increased with increasing filler loading compared to the pure polymer membrane.[79]

MMMs were also fabricated using spin coating, achieving significant improvements over the neat polymer membranes with NH_2-MIL-53, ZIF-8.[105] Mixed matrix membranes were fabricated by dispersing the 100 nm ZIF-71 nanoparticles incorporated in 6FDA-Durene polyimide membranes.[106] The incorporation of ZIF-71 nano-particles improves both gas permeation properties and membrane resistance against plasticization. The MMM embedded with 20 wt% ZIF-71 also displays a remarkable enhancement of C_3H_6 permeability from 57.6 Barrer to 371 Barrer without significantly compromising the ideal C_3H_6/C_3H_8 selectivity.[106] A new kind of self-supported dual layer mixed matrix membrane was developed using ZIF-8 as inorganic filler in PEBAX-2533 polymer matrix.[100] There is an inorganic base layer closely connected through a network of polymer and one polymer base layer with a few inorganic particles dispersed and the two layers cannot be separated by peeling off. This morphology is a different morphology of inorganic–organic matrix membranes compared to the common ones in which inorganic fillers are more homogenously dispersed in the polymer matrix. The gas permeability increased as the inorganic filler content increased in these matrix membranes.[100] The fabrication of dense mixed matrix membranes was also extended to scalable hollow fiber geometry. Dual layer ZIF-8/6FDA-DAM mixed-matrix hollow fiber membranes were successfully fabricated with ZIF-8 nanoparticle loading up to 30 wt% using the conventional dry-jet/wet-quench fiber spinning technique.[107] These mixed-matrix hollow fibers showed significantly enhanced C_3H_6/C_3H_8 selectivity ($\sim$28) that was consistent with mixed-matrix dense films.

Overall, MMM possess great potential for tremendous improvements in both permeability and selectivity especially for metal-organic frameworks and porous organic frameworks as an additive. The separation factor in MMM depends on polymer-filler interface and nature of filler dispersion. The next generation MMM can be produced using nano-size fillers without agglomeration. The nano-sized particles would have more uniform particle distribution in polymer matrix than micro-size particles and allow forming thin films for high throughput membrane development. The transformation of separation performance of film-based MMMs to the industry relevant hollow fiber-based MMMs is highly critical and difficult due to different preparation conditions used in both flat and hollow fiber membrane forms. Further research and development are needed for new additives synthesis, polymer selection and membrane preparation techniques.

All the experimental results concluded that MOFs are better compatible with the polymers than other inorganic filler particles likes zeolites, carbon molecular sieves, metal nanoparticles, etc. and showed great improvement in gas separation properties than the pure polymer membranes. Most importantly, there is a large library of MOF structures reported in the literature with different pore structures and chemical functionalities, which provides the opportunity to prepare high number of mixed matrix membranes. It is very important to use the theoretical modeling and computational screening techniques to investigate the potential polymer-MOF pairs to achieve high gas separation performances instead of experimentally studying these unlimited MMM structures. Mixed matrix membranes (MMMs) with MOFs provide potential alternatives as enhanced gas separation membranes and have a lot of room for improvement of the performance.

Pure MOF membranes showed very high gas permeabilities, around 1–3 orders of magnitude, compared to pure polymeric membrane.[12] Interestingly, dispersing these highly permeable inorganic particles in the polymers did not show very high improvement in gas transport properties. The overall gas separation performance is still close to polymer membranes even after 30–40 wt% loadings of inorganic particles. Moore *et al.* found that the gas permeability of MMM increases significantly compared to pure polymer membranes if the permeabilities of the faster permeating component through the polymer and sieve are approximately matched.[13] If polymer permeability is much less than the filler permeability, the resultant MMM doesn't show great improvement in gas permeability. Recently, metal-organic framework-based MMM gained lots of interest and promise. MOF-5, one of the well-known MOF showed intrinsic CO_2 permeability of 4,354 Barrer with very low selectivity.[110] Addition of MOF-5 in Matrimid showed improvement in both CO_2 permeability and CO_2/N_2 selectivity.[111] Similar to SAPO-34 and CMS, the increase in the gas separation performance is very minimal. There is a need to develop the techniques to fabricate these MMM with higher volume loadings of fillers in order to achieve the gas separation performance close to that of inorganic fillers. It is a challenge to increase the volume loading above optimum loading because of the formation of defects and weak mechanical properties of MMM.

9.4. Gas separation performance calculations of MMM using predictive models

The gas separation performance of the MMMs has been influenced by many factors such as gas species, percentage of filler loading, pore size of fillers, flexibility of the filler structure, polymer glass transition temperature

(Tg), fractional free volume (FFV) of polymers, dispersion of the fillers, and polymer/filler interface morphology.[14,19,21,27,33,38,69,112–115] Essentially, these properties are determined by the interactions among gas molecules, polymers and fillers, which can be computed using molecular simulation with verified force fields accurately. Molecular modeling can reach tens nanoseconds and nanometers at maximum, which is large enough to compute the gas transportation in the homogenous phases like pure polymer membranes and inorganic porous fillers. However, the size of fillers in the MMMs is larger than 100 nanometers mostly.[21] A representative "unit" of hybrid MMMs is significantly beyond the scope of molecular modeling and cannot be simulated by molecular simulation alone.

Permeation models such as Maxwell model, Bruggeman model, Lewis-Nielson model and Pal model have been applied to describe systems at larger scales and predict the performance of the MMMs.[14] To make the prediction, the gas permeabilities in dispersed phases and continuous phases need to be known as inputs for permeation model. The experimental data of gas permeation in polymer membranes have been reported comprehensively.[1,2] However, it is still hard to measure the gas permeability in inorganic particles using experiments. Although the gas permeability of inorganic films have been studied in some experimental research groups,[9–12] the gas permeabilities in the particles used in MMMs are much different from inorganic membranes because of the grain boundary defects between the crystals and other defects in the membrane which was formed by many crystals.[33] Molecular simulation is an important tool to investigate the gas transportation in single phase, especially in filler crystals.[116] The interfacial structure between fillers and polymers influences the performance of composites significantly. It is hard to explore the filler/polymer interface at nanoscale using current experimental instruments or permeation models. Molecular modeling is an ideal method to fill this gap. Therefore, it is necessary to combine molecular models with permeation models in order to predict the gas separation performance of MMMs.

9.4.1. *Permeation Models of MMMs with ideal particle-filler interface*

Most of the permeation models have been developed with the assumption of ideal interfacial morphology between the particle and polymer. In an ideal interfacial morphology, there are no defects, densifications and distortions in the membranes (Case I in Figure 9.3). Under such consideration, the properties of polymers and fillers at interfacial region are the same as in their bulks. A well-known permeation model is Maxwell-Wagner-Sillars equation that was originally developed from the Maxwell's theory for predicting the electrical conductivity of hybrid materials.[117–119] It has been applied to calculate permeability of the MMMs because of the similarity between electrical conduction through composite materials and gas permeation through MMMs.

The full version of Maxwell-Wagner-Sillars equation is:

$$P_{MMM} = P_c \left[\frac{nP_d + (1-n)P_c - (1-n)\phi_d(P_c - P_d)}{nP_d + (1-n)P_c + n\phi_d(P_c - P_d)} \right]$$

where P_{MMM} is the mixed-matrix membrane permeability, P_d is the permeability of the dispersed filler phase and P_c is the permeability of the continuous polymer phase, and ϕ_d is the volume fraction of the fillers. $0 < n < 1$ is the shape factor of fillers, at the limits of $n = 0$ for the parallel platelets, which represents an arithmetic average of gas permeation of fillers and polymers and $n = 1$ for gas transportation through the laminates when the equation becomes Hamilton-Crosser model for cylindrical fillers (e.g. carbon nanotunbes). When $n = 1/3$ for the spherical fillers, the Maxwell-Wagner-Sillars equation can be simplified into the most well-known and commonly used Maxwell equation:

$$P_{MMM} = P_c \left[\frac{P_d + 2P_c - 2\phi_d(P_c - P_d)}{P_d + 2P_c + \phi_d(P_c - P_d)} \right]$$

Although the maximum packing volume fraction can reach up to 0.64 for random close packing of uniform spherical fillers, the Maxwell-type equations can only be applied to the systems for the low volume loading of fillers in MMMs because the model was derived on the condition of streamline morphology around fillers, which can be interfered and even broken by the close presence of fillers. In practice, over 0.3 volume loading of fillers can lead to significant errors, using Maxwell models to predict gas permeation through the MMMs.

For the systems of higher volume loading of fillers, other models have been developed. For example, Bruggeman equation,[120] adopted from the model of dielectric constant in composite materials, has been applied to calculate the gas permeation through mixed matrix membranes as shown:

$$\frac{P_{MMM}}{P_c} = \frac{1}{(1-\phi_d)^3}\left(\frac{\frac{P_{MMM}}{P_c} - \frac{P_d}{P_c}}{1 - \frac{P_d}{P_c}}\right)^3$$

However, the prediction capability of Bruggeman model cannot reach the maximum volume loading. The exclusion of factors of filler shape, size distribution and aggregation of fillers decrease the accuracy of the prediction. Bruggeman equation is not feasible to use for the prediction of gas separation performance because of implicit functional form.

The Lewis-Nielsen model is the one that can be applied to the system with loading at maximum packing volume fraction of fillers.[121,122] It has been adopted from the field of material mechanics to calculate the elastic modulus of composites, as expressed as below:

$$P_{MMM} = P_c\left[\frac{1 + 2\phi_d\frac{\left(\frac{P_d}{P_c}\right)-1}{\left(\frac{P_d}{P_c}\right)+2}}{1 - \psi\phi_d\frac{\left(\frac{P_d}{P_c}\right)-1}{\left(\frac{P_d}{P_c}\right)+2}}\right],$$

where, $\psi = 1+\phi_d\left(\frac{1-\phi_m}{\phi_m^2}\right)$ and ϕ_m is the maximum packing volume fraction of fillers, which is a function of filler size, filler distribution and aggregation. The Lewis–Nielsen model includes the effects of morphology partially on permeability via setting up ϕ_m. With $\phi_m = 0.64$, the random close packing of uniform spheres has been assumed in many systems. When $\phi_m \rightarrow 1$, the Lewis–Nielsen model reduces to Maxwell model.

Pal model is another permeation model that can be applied to the system with filler loading at maximum packing volume fraction as shown[115,123]:

$$\left(\frac{P_{MMM}}{P_c}\right)^{\frac{1}{3}}\left(\frac{\frac{P_d}{P_c}-1}{\frac{P_d}{P_c}-\frac{P_{MMM}}{P_c}}\right) = \left(1-\frac{\phi_d}{\phi_m}\right)^{-\phi_m}$$

The Pal model is an implicit function, and reduces to Bruggeman equation when $\phi_m \rightarrow 1$.

Among the above-mentioned models, Maxwell model is the simplest one. In comparison to the others, Maxwell model cannot be applied in the high filler loading systems. As shown in Figure 9.9,[21] under low fractional filler loading (<0.2), the simulation results of Maxwell and Bruggeman

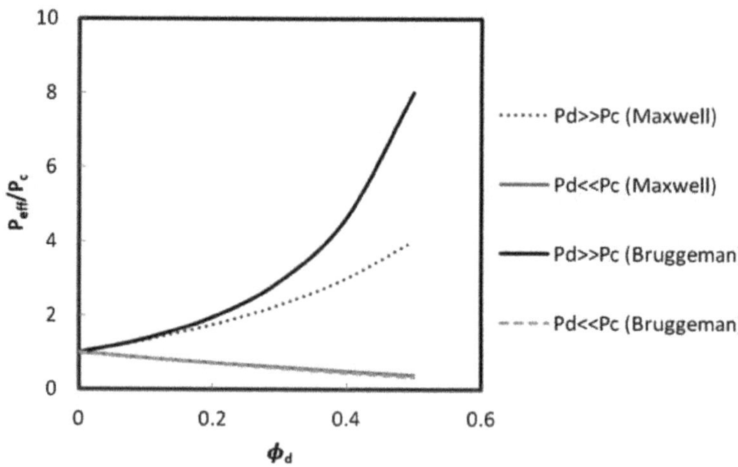

Figure 9.9. Simulation of the permeability of an ideal mixed matrix membrane as a function of volume fraction of the inorganic particles.[21] (Reproduced from Ref. 21 with permission from the Royal Society of Chemistry.)

equations are almost identical since chain conformation of polymers around fillers are not affected by the neighboring fillers and kept the same as in polymer bulk. At a higher fractional particle loading (>0.2), there is a substantial deviation in the simulation results between the Maxwell and Bruggeman models, and Maxwell model is no longer valid at this loading since polymer chains between the neighboring fillers distort seriously. Nevertheless, Maxwell equation is still the most commonly used permeation model for the following reasons: (1) it is very straightforward and convenient to use, and (2) in practice, the filler volume loading is usually lower than 0.3 in most of MMMs since it is difficult to fabricate defect-free membranes at high filler loading systems. The above discussion is based on the assumption of ideal interface between fillers and polymers. Once defects are formed in the MMMs (as described in Figure 9.3), a modification for the original equations has to be applied.

9.4.2. *Permeation Models of MMMs with non-ideal particle-filler interface*

It is helpful to avoid the complexity in building permeation models for MMMs with the assumption of ideal interface. However, these "ideal" models could not describe the real MMMs accurately if non-ideal interface exists. As shown in Figure 9.3, the interfacial defects can be categorized into five cases as presented by Moore and Koros.:[13] I) rigidified polymer layers wrapped around the fillers, II) large voids ($>\sim$10–100 nm) at the polymer/filler interface, which could be an unfavorable result after the MMM processing, III) small voids at the interface mainly because of the rigidity of polymer chains and less attractive interaction between polymers and fillers, IV) clogged fillers by gas molecules because of strong holding from the inside surface of fillers, and V) bottlenecked at filler side of the interfacial region.

Moore and Koros attributed the formation of interface cases I–III to the stresses encountered during the MMM fabrication. In case I, gas permeability is decreased at the rigidified polymers wrapping around the filler surface. The rigidified polymer layer has been described to be a denser and less mobile phase because of rigid structure of fillers and contraction of polymers with long time relaxation after solvent evaporation. The rigidified polymers are similar to semi-crystal polymers, in which the gas permeability is 1/3 to 1/4 of it in amorphous polymers. As shown in Figure 9.3, when the MMMs have been fabricated as films on substrates, the contraction of polymers is asymmetric and results in void structures at the filler/polymer interface. Moore and Koros classified case III as a special situation with small interfacial gap of case II. However, even if the interface stress has been relaxed, the interfacial defects cannot be eliminated completely. The interaction between polymers and fillers also contributes to the formation of case I and case III. When the interaction between polymers and fillers is attractive, it drives the polymers into a dense rigidified layer. In contrast, the repulsive interaction between polymers and fillers makes nano-size voids at interface. Since the molecular interaction is nano-distance force, the preference between polymers and fillers has no effects on void formation in case II with large size (e.g. $>$10–100 nm). In some contexts, rigidified layers have been described as densified layers, immobilized structure, *pancake* morphology, etc. Actually, the *pancake* morphology of polymer chains always exist in the region close to the filler surface whatever the surface is attractive, neutral or repulsive due to the entropy effects. As well, these chains lose mobility because of the deformation and blockage by filler surface. At attractive filler surface, the density at rigidified layer is higher than in bulk and the polymer density at repulsive filler surface is lower than in the bulk.

Moore and Koros modified the basic Maxwell equation to be applicable for the MMM systems involving non-ideal interfaces as shown in Table 9.2[13] The Maxwell equation has been first applied to the dispersed filler and interface layer of voids assumed to involve Knudsen diffusion, rigidification with permeation reduction by *chain immobilization factor* β or *permeability reduction factor* β. Then, the interface layer and fillers together have been joined as a pseudodispersed phase. The Maxwell equation has been applied the second time to the pseudodispersed phase and matrix. The modified Maxwell equation has included interface defects quantitatively, but still has been subjected to the same limitation as original Maxwell equation in filler loading, particle size distribution, particle shape, and particle aggregation. There are some more complicated situations in MMMs involving more than one interfaces: e.g. polymer rigidification happens around the outside of the fillers with pore blockage occurring inside of the particle, then the basic Maxwell equation has to be applied three times to describe the overall MMM performance.[13] Similar strategy of *nested application* has also been employed in other models.

Felske developed a model originally for thermal conductivity of composites of core-shell fillers (core filler covered with an interfacial layer), which is also valid for permeability prediction. The Felske model can be expressed as below[124]:

$$P_r = \frac{P_{eff}}{P_c} = \frac{2(1-\phi_d) + (1+2\phi_d)\beta/\gamma}{(2+\phi_d) + (1-\phi_d)\beta/\gamma}$$

where, $\beta = (2+\delta^3)\lambda_{dm} - 2(1-\delta^3)\lambda_{Im}$

$$\gamma = (1+\delta^3) - (1-\delta^3)\lambda_{dI}$$

where, δ is the ratio of outer radius of interfacial shell to a core radius. $\lambda_{dm} = \frac{P_d}{P_m}$, $\lambda_{Im} = \frac{P_I}{P_m}$, and $\lambda_{dI} = \frac{P_d}{P_I}$ are ratios of permeabilities in dispersed fillers, polymer matrix and interphase individually. When $\delta = 1$, the interfacial layer is absent and the Felske model reduces to the original Maxwell model.

Felske model provides almost the same predictions as the (modified) Maxwell model can do, and suffers the same limitations (e.g. $\phi_d < 0.2$) as the Maxwell model. Pal further modified the Felske model by introducing composite morphology parameter ψ, a function of ϕ_m (maximum packing

Table 9.2. Comparison of models for cases I, II and III, and IV based on nested applications of the Maxwell model.[13]

	Case I, matrix rigidification	Case II-III, sieve-in-cage	Case IV-V, 'clogged' or 'bottle-bottlenecked' zeolites
Parameters	β, 'Chain immobilization factor' $P_I = P_c/\beta$ l_I, Thickness of the rigidified region	P_I, Void permeability $P_I = DS$ $D^a = D_{Knudsen}\left(1 - \frac{\sigma_p}{2l_I}\right)$ $S^a = \frac{1}{RT}\left(1 - \frac{\sigma_p}{2l_I}\right)^2$ l_I, Void thickness	β', 'Permeability reduction factor' $P_I = P_c/\beta'$ l'_I, Thickness of the reduction permeability region
Model for the pseudo-dispersed phase permeability	$P_{eff} = P_I\left[\frac{P_d+2P_I-2\phi_s(P_I-P_d)}{P_d+2P_I+\phi_s(P_I-P_d)}\right]$ where $\phi_s = \frac{\phi_d}{\phi_d+\phi_I} = \frac{r_d^3}{(r_d+l_I)^3}$ or $\frac{(r_d-l'_I)^3}{r_d^3}$		
Pseudo-dispersed phase components	Zeolite + surrounding polymer of reduced permeability	Zeolite + surrounding void	Zeolite (including bulk zeolite and region of reduced permeability)
Model for the overall mixed matrix material	$P_{3mm} = P_c\left[\frac{P_{eff}+2P_c-2(\phi_d+\phi_I)(P_c-P_{eff})}{P_{eff}+2P_c+(\phi_d+\phi_I)(P_c-P_{eff})}\right]$		

[a]The term $\left(1 - \frac{\sigma_p}{2l_I}\right)$ must be added to account for the finite size of the penetrants when the void dimension is of the same order of magnitude as the penetrants. $D_{Knudsen}$ is given by the expression: $D_{Knudsen} = (d_{Pore}/3)(8RT/\pi M_p)^{1/2}$, which is based on the kinetic theory of gases. We used the hydraulic diameter of the pore ($\sim 2l_I$) in place of the diameter of the pore, d_{pore}. P_I, P_c and P_d are gas permeabilities at interface, polymer and particle separately. P_{eff} and P_{3mm} are gas permeabilities at the pseudo-dispersed phase and the whole membrane individually. D and S are gas diffusivity and solubility in the voids separately. $D_{Knudsen}$ is the Knudsen diffusivity. σ_P and M_p are dynamical dimeter and mass weight of the penetrate (gas) molecule. ϕ_d and ϕ_I are volume fractions of dispersed phase and interface phase individually. r_d is the radius of particle.

volume fraction of core-shell filler) and ϕ_d as defined in Equation 5.[115] The loading range of modified Felske model has been extended to $\rightarrow \phi_d\phi_m$. When $\phi_m = 1$, the modified Felske model reduces to original one. When $\delta = 1$, the modified Felske model reduces to Lewis-Nielsen model. And when both ϕ_m and δ are equal to 1, the modified Felske model reduces to original Maxwell model.

Shimekit *et al.* modified Pal model by including the effects of interfacial rigidification on the permeability of MMMs as below[125]:

$$P_r = f(\delta, \lambda_{Im}, \lambda_{dI}, \phi_d, \phi_m, \phi_s)$$

where δ, λ_{Im}, $\lambda_{dI}\phi_d$, ϕ_m and ϕ_s are defined as the same as in other models discussed in the context.

The Pal model for relative permeability P_r can be rewritten as follows:

$$(P_r)^{1/3}\left(\frac{\lambda_{dI} - 1}{\lambda_{dI} - P_r}\right) = \left(1 - \frac{\phi_d}{\phi_m}\right)^{-\phi_m} \tag{12a}$$

by replacing P_r with $\frac{P_{eff}}{P_I}$, the effective permeability for pseudodispersed phase (the combination of filler and interphase) can be calculated:

$$\left(\frac{P_{eff}}{P_I}\right)^{1/3}\left(\frac{\lambda_{dI} - 1}{\lambda_{dI} - \frac{P_{eff}}{P_I}}\right) = \left(1 - \frac{\phi_d}{\phi_m}\right)^{-\phi_m} \tag{12b}$$

As in the modified Maxwell model, the permeability at rigidified interphase is defined as $P_I = \frac{P_m}{\beta}$, and β the chain immobilization factor is usually set to 3.

In all the models for non-ideal MMMs, δ, the ratio of outer radius of interfacial shell to a core radius or the thickness of interphase has to be determined. However, it cannot be measured directly using experimental ways. Alternately, fitting program with the least-square method has been used to estimate the value of δ. In this procedure, experimental permeation data need to be used for the fitting. This actually fails the capability of permeation models to predict the performance. Therefore, independent tools to determine interphase thickness are desired.

9.4.3. *Molecular models*

Non-ideal interface happens in almost all MMMs and is critical to affecting the gas separation performance of MMMs, but interfacial information of filler/polymer is hard to be measured using experimental tools. A representative unit of MMMs involving three phases (bulk polymer, interface and a whole filler) is too large to be simulated using molecular modeling. Under current computing resource, using molecular modeling to investigate local interface is a feasible way and the most important part to understand MMMs.

Molecular simulation can be divided into three main categories by length/time scales: atomistic model, coarse-grained off-lattice model and coarse-grained lattice model. Atomistic simulation is the most accurate, but most time consuming; coarse-grained lattice model is the least time consuming and provides general trends, but lose chemical details; in comparison, the off-lattice model is a compromise solution between the atomistic model and coarse-grained lattice model. It takes a very long time to relax polymer chains and the relaxation time (τ) increases nonlinearly with the chain length in condensed phase; e.g. τ of a chain with 10 segments is about 10 ns and $\tau \propto$ L3-3.4.[122]

To complete a simulation in a reasonable time, most work using atomistic model starts from a desired equilibrium state that is determined by molecular mechanics.[126] However, it is difficult to apply this strategy to the systems of condensed polymers on inhomogeneous surfaces like filler surfaces in MMMs. In contrast, many polymer microstructures confined by surface have been predicted using coarse-grained lattice models and verified by experiments.[127–132] Employing lattice Monte Carlo simulation, Termonia predicted the stretching and widening of polymer chains near a nanoparticle surface and form a 1–2 nm interfacial region.[133] When the two small particles approach in a distance smaller than the radius of gyration of the polymer chain, an overall swelling of polymer chains between the particles have been observed. Gao *et al.* observed the formation of the pancake-like morphology of the polymer in the region close to a nanosheet surface using coarse-grained molecular dynamics simulations.[134]

It is interesting that this phenomena is independent of the interfacial strength since configurational entropy rather than interaction energy dominates the interfacial morphology. However, the interfacial interaction strength still influences the glass transition temperature. It has been found that the glass transition temperature for repulsive interface interaction is lower, and for attractive interface interaction is higher than the pure polymer system. Furthermore, they predicted the rigidification of the polymer

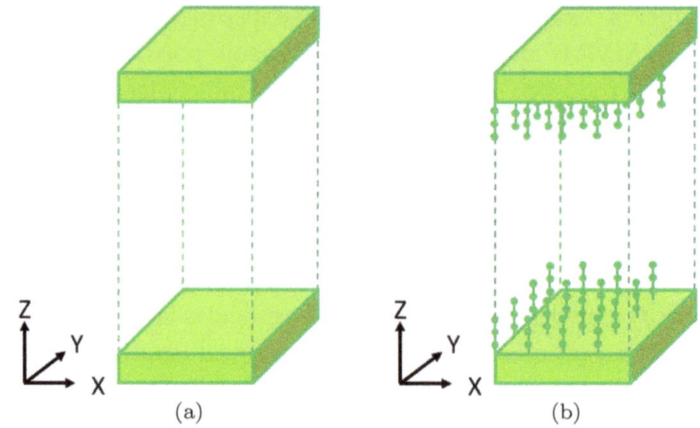

Figure 9.10. Schematic representation of the confined systems with two parallel surfaces. (Reprinted from Ref. 135 with permission from Elsevier.)[135]

layer on a strongly attractive surface. Very recently, Feng *et al.* studied the polymer morphology and free volume at polymer/particle interface using lattice coarse-grained Monte Carlo simulations as shown in Figure 9.10.[135] It was found that the interfacial properties are dominated by the interaction of the polymer with the filler particle and the geometry of the filler particle surface. Filler particles that have brush-like surfaces tend to result in larger free volume at the interface than occurs with smooth surfaced filler particles. Until now, the research on MMMs using molecular simulation is still rare and incomplete. In future, it is desired to develop: (1) Accurate force fields for coarse-grained model in order to simulate the polymer/filler interface quantitatively, and (2) novel method for predicting the gas permeability at interface region.

Budhathoki *et al.* used computational methods to predict the gas transport properties of the MOF based mixed matrix membranes.[136] Atomostic simulations were used to calculate the gas adsorption and permeability in 112,888 MOFs. Widom particle insertion method was used to calculate the Henry's law constants of CO_2 and N_2 of all these MOFs and self-diffusivities of CO_2 and N_2 in all these MOFs were also computed using the molecular dynamics code LAMMPS50 at 298 K. The CO_2/N_2 gas transport properties of over a million hypothetical MMMs with combination of 9 different polymers and 112,888 MOFs were predicted using the Maxwell model. The gas transport properties of these 9 polymers were measured experimentally and the gas transport properties of the 112,888 MOFs were computed. Ideal CO_2/N_2 selectivities were also calculated by taking the ratio of the predicted pure gas permeabilities of the MMMs. The computed gas transport properties of the MMMs were also compared with the experimentally measured properties as shown in the Figure 9.11. The computed properties are in reasonably close to the experimentally measured values showing that the methods used to predict the gas transport properties of MOF based MMM are reliable. Figure 9.12 shows the predicted CO_2 permeability and CO_2/N_2 selectivity of several MMM in comparison with the pure polymer properties. As shown in the Figure 9.12, the gas transport properties of the polymers increased with the addition of MOFs and in some cases it was decreased. It was observed that polymer with high gas permeabilities such as polymers of intrinsic microporosity (PIM) and poly[1-trimethylsilyl-1-propyne] (PTMSP) has great improvement with the addition of MOF particles, whereas lower permeable such as Matrimid based MMM did not show

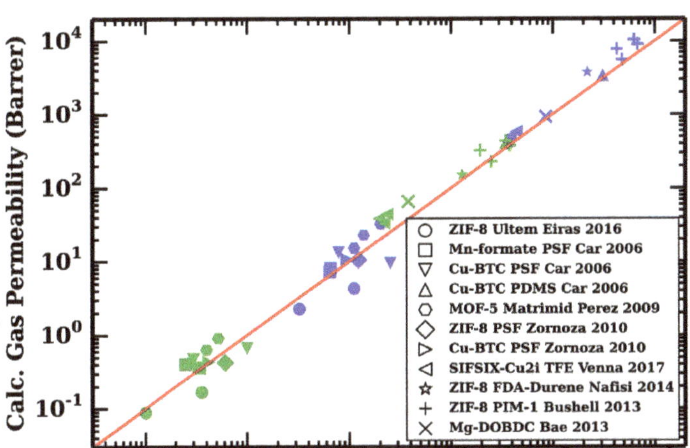

Figure 9.11. Selectivity as a function of CO_2 permeability for MMMs based on a) the CoRE database and b) the hypothetical MOF database. The diagonal line is the Robeson bound. The square symbols mark the selectivity and permeability of the neat polymers and the small round symbols mark the predicted selectivity and permeability of the MMMs based on each of the nine polymers. (Reproduced from Ref. 21 with permission from the Royal Society of Chemistry.)

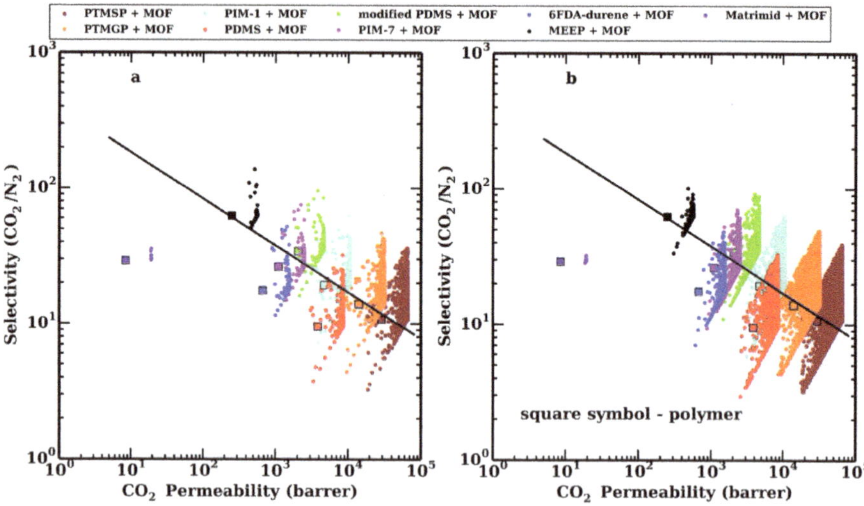

Figure 9.12. Predicted versus measured CO_2 and N_2 permeabilities of mixedmatrix membranes using Maxwell model with the experimental values. Blue color represents CO2permeability and green color represents N_2 permeability. The red diagonal line indicates calculated permeability = experimental permeability. (Reproduced from Ref. 21 with permission from the Royal Society of Chemistry.)

significant improvement. In order to show the highest improvement in gas transport properties in MMMs, the polymer should have very high permeability and it needs to paired with a MOF that has at least 100 times more gas permeability and 1000 times higher selectivity compared to the polymer.

9.5. Final remarks on MOF-based MMM

Zeolites, silica and carbon molecular sieves were used as filler in the initial phases of MMM research. But it gained more interest among the researchers after the invention of other porous materials such as metal-organic frameworks, carbon nanotubes, covalent organic frameworks and porous polymers. Overall, MMM have been studied widely in the literature in order to optimize the interface structure, size of particles, compatibility between polymer and filler, mechanical properties etc. using a variety of inorganic materials. There are very few reports on studying the effect of morphology and size of the above-mentioned filler particles on the gas transport properties. Further research on understanding the size and morphology effect including the alignment of the pores will be useful for the further development of these MMM. Agglomeration and sedimentation of the particles also poses a challenge to process the MMM, especially using nanomaterials. Synthesizing the particles during the fabrication of membranes also results in good dispersion and formation of smaller particles to improve gas separation performances. Future research should also extend on developing novel techniques to keep these particles in the dispersed phase.

Several reports were published on the tuning of the interface using ionic liquids, plasticizers, surface functionalization and by applying several processing modifications. But the complete interface structure and its involvement in MMM is still unknown. The research must be focused on understanding the complex structure of the MMM especially at the interface of polymer and filler in order to develop them further. Several researchers used conventional characterization techniques such as SEM, TEM, CT scan and PALS to study the limited interface volume, but all those methods give qualitative information. Experimental techniques needs to be developed to study the interface structure and its importance in improving the gas transport properties. The research should also focus on developing molecular simulation models to analyze the polymer-filler interface quantitatively. It is also important to study novel methods for predicting the gas transport properties of MMM including the non-ideal effects. Several mathematical models were developed for the prediction of the gas transport properties of the MMM, but these methods were based on several assumptions and difficult to predict the properties accurately. This also helps in the proper selection of materials using a consistent approach.

Although MMM shows gas separation performance well above the Robeson upper bound, economic processing of the MMM at industrial scale is the main hurdle for applying these membrane in the industry. The transformation of separation performance of film-based MMMs to the industry in relevant hollow fiber-based MMMs is highly critical and difficult due to challenging hollow fiber preparation conditions. Hollow fiber MMM are advantageous at industrial scale compared to flat sheet MMM because of their high packing density. Research on the fabrication of mixed matrix hollow fiber membranes is very sporadic. The research needs more focus on studying the single layer and dual layer hollow fiber mixed matrix membranes. Fabrication of thin mixed matrix membrane on the hollow fiber supports is also a challenge. Synthesis of smaller porous particles is advantageous and essential to prepare thin MMM and the research should also concentrate on the development of techniques to keep these particles dispersed. These smaller particles also provide more interfacial area and disrupt the polymer chain packing which enhances the gas transport properties of MMM.

References

1. Robeson, L. M. Correlation of separation factor versus permeability for polymeric membranes. *J. Memb. Sci.* **62**, 165–185 (1991).
2. Robeson, L. M. The upper bound revisited. *J. Memb. Sci.* **320**, 390–400 (2008).
3. Freeman, B. D. Basis of permeability/selectivity tradeoff relations in polymeric gas separation membranes. *Macromolecules* **32**, 375–380 (1999).
4. Carreon, M. A., Li, S., Falconer, J. L. & Noble, R. D. SAPO-34 Seeds and membranes prepared using multiple structure directing agents. *Adv. Mater.* **20**, 729–732 (2008).
5. Carreon, M. L., Li, S. & Carreon, M. A. AlPO-18 membranes for CO_2/CH_4 separation. *Chem. Commun. (Camb).* **48**, 2310–2 (2012).
6. Carreon, M. A., Li, S., Falconer, J. L. & Noble, R. D. Alumina-supported SAPO-34 membranes for CO_2/CH_4 separation. *J. Am. Chem. Soc.* **130**, 5412–3 (2008).
7. Li, S. & Fan, C. Q. High-flux SAPO-34 membrane for CO_2/N_2 separation. *Ind. Eng. Chem. Res.* **49**, 4399–4404 (2010).
8. Li, S. *et al.* Scale-up of SAPO-34 membranes for CO_2/CH_4 separation. *J. Memb. Sci.* **352**, 7–13 (2010).
9. Venna, S. R., Zhu, M., Li, S. & Carreon, M. A. Knudsen diffusion through ZIF-8 membranes synthesized by secondary seeded growth. *J. Porous Mater.* **21**, 235–240 (2013).
10. Venna, S. R. & Carreon, M. A. Highly permeable zeolite imidazolate framework-8 membranes for CO_2/CH_4 separation. *J. Am. Chem. Soc.* **132**, 76–8 (2010).
11. Venna, S. R. & Carreon, M. A. Amino-functionalized SAPO-34 membranes for CO_2/CH_4 and CO_2/N_2 Separation. *Langmuir* **27**, 2888–94 (2011).
12. Venna, S. R. & Carreon, M. A. Metal organic framework membranes for carbon dioxide separation. *Chem. Eng. Sci.* **124**, 3–19 (2015).
13. Moore, T. T. & Koros, W. J. Non-ideal effects in organic–inorganic materials for gas separation membranes. *J. Mol. Struct.* **739**, 87–98 (2005).
14. Vinh-Thang, H. & Kaliaguine, S. Predictive models for mixed-matrix membrane performance: A review. *Chem. Rev.* **113**, 4980–5028 (2013).
15. Vinoba, M. *et al.* Recent progress of fillers in mixed matrix membranes for CO_2 separation: A review. *Sep. Purif. Technol.* **188**, 431–450 (2017).
16. Nik, O. G., Chen, X. Y. & Kaliaguine, S. Functionalized metal organic framework-polyimide mixed matrix membranes for CO_2/CH_4 separation. *J. Memb. Sci.* **413**, 48–61 (2012).

17. Liu, Y., Zeng, G., Pan, Y. & Lai, Z. Synthesis of highly c-oriented ZIF-69 membranes by secondary growth and their gas permeation properties. *J. Memb. Sci.* **379**, 46–51 (2011).
18. Caro, J. Are MOF membranes better in gas separation than those made of zeolites? *Curr. Opin. Chem. Eng.* **1**, 77–83 (2011).
19. Chung, T.-S., Jiang, L. Y., Li, Y. & Kulprathipanja, S. Mixed matrix membranes (MMMs) comprising organic polymers with dispersed inorganic fillers for gas separation. *Prog. Polym. Sci.* **32**, 483–507 (2007).
20. Tanh Jeazet, H. B. *et al.* Metal-organic frameworks in mixed-matrix membranes for gas separation. *Dalt. Trans.* **41**, 14003 (2012).
21. Dong, G. *et al.* Challenges and opportunities for mixed-matrix membranes for gas separation. *J. Mater. Chem. A* **1**, 4610 (2013).
22. Baker, R. W. *Membrane Technology and Applications*. (John Wiley & Sons, Ltd, 2004). doi:10.1002/0470020393.
23. Loeb, S. & Sourirajan, S. Sea water demineralization by means of an osmotic membrane. in *Advances in Chemistry, Vol. 38* 117–132 (1963). doi:10.1021/ba-1963-0038.ch009.
24. Merkel, T. C. *et al.* Ultrapermeable, reverse-selective nanocomposite membranes. *Science* **296**, 519–22 (2002).
25. Jiang, L. Y., Chung, T. S., Cao, C., Huang, Z. & Kulprathipanja, S. Fundamental understanding of nano-sized zeolite distribution in the formation of the mixed matrix single- and dual-layer asymmetric hollow fiber membranes. *J. Memb. Sci.* **252**, 89–100 (2005).
26. Jiang, L. Y., Chung, T. S. & Kulprathipanja, S. An investigation to revitalize the separation performance of hollow fibers with a thin mixed matrix composite skin for gas separation. *J. Memb. Sci.* **276**, 113–125 (2006).
27. Mahajan, R., Burns, R., Schaeffer, M. & Koros, W. J. Challenges in forming successful mixed matrix membranes with rigid polymeric materials. *J. Appl. Polym. Sci.* **86**, 881–890 (2002).
28. Pechar, T. W. *et al.* Fabrication and characterization of polyimide–zeolite L mixed matrix membranes for gas separations. *J. Memb. Sci.* **277**, 195–202 (2006).
29. Ismail, A. F., Kusworo, T. D. & Mustafa, A. Enhanced gas permeation performance of polyethersulfone mixed matrix hollow fiber membranes using novel Dynasylan Ameo silane agent. *J. Memb. Sci.* **319**, 306–312 (2008).
30. Gür, T. M. Permselectivity of zeolite filled polysulfone gas separation membranes. *J. Memb. Sci.* **93**, 283–289 (1994).
31. Kim, S., Pechar, T. W. & Marand, E. Poly(imide siloxane) and carbon nanotube mixed matrix membranes for gas separation. *Desalination* **192**, 330–339 (2006).
32. Funk, C. V. & Lloyd, D. R. Zeolite-filled microporous mixed matrix (ZeoTIPS) membranes: Prediction of gas separation performance. *J. Memb. Sci.* **313**, 224–231 (2008).
33. Adatoz, E., Avci, A. K. & Keskin, S. Opportunities and challenges of MOF-based membranes in gas separations. *Sep. Purif. Technol.* **152**, 207–237 (2015).
34. Ahn, J., Chung, W.-J., Pinnau, I. & Guiver, M. D. Polysulfone/silica nanoparticle mixed-matrix membranes for gas separation. *J. Memb. Sci.* **314**, 123–133 (2008).

35. Anson, M., Marchese, J., Garis, E., Ochoa, N. & Pagliero, C. ABS copolymer-activated carbon mixed matrix membranes for CO_2/CH_4 separation. *J. Memb. Sci.* **243**, 19–28 (2004).
36. Bastani, D., Esmaeili, N. & Asadollahi, M. Polymeric mixed matrix membranes containing zeolites as a filler for gas separation applications: A review. *J. Ind. Eng. Chem.* **19**, 375–393 (2013).
37. Bertelle, S., Gupta, T., Roizard, D., Vallières, C. & Favre, E. Study of polymer-carbon mixed matrix membranes for CO_2 separation from flue gas. *Desalination* **199**, 401–402 (2006).
38. Goh, P. S., Ismail, A. F., Sanip, S. M., Ng, B. C. & Aziz, M. Recent advances of inorganic fillers in mixed matrix membrane for gas separation. *Sep. Purif. Technol.* **81**, 243–264 (2011).
39. Vu, D. Q., Koros, W. J. & Miller, S. J. Mixed matrix membranes using carbon molecular sieves. *J. Memb. Sci.* **211**, 335–348 (2003).
40. Zhao, D. *et al.* Effect of graphene oxide on the behavior of poly(amide-6-b-ethylene oxide)/graphene oxide mixed-matrix membranes in the permeation process. *J. Appl. Polym. Sci.* **132**, n/a-n/a (2015).
41. Moore, T. T., Mahajan, R., Vu, D. Q. & Koros, W. J. Hybrid membrane materials comprising organic polymers with rigid dispersed phases. *AIChE J.* **50**, 311–321 (2004).
42. Tantekin-Ersolmaz, Ş. B. *et al.* Effect of zeolite particle size on the performance of polymer–zeolite mixed matrix membranes. *Journal of Membrane Science* **175**, (2000).
43. Su, N. C., Smith, Z. P., Freeman, B. D. & Urban, J. J. Size-Dependent Permeability Deviations from Maxwell's Model in Hybrid Cross-Linked Poly(ethylene glycol)/Silica Nanoparticle Membranes. (2015).
44. Ordoñez, M. J. C., Balkus, K. J., Ferraris, J. P. & Musselman, I. H. Molecular sieving realized with ZIF-8/Matrimid® mixed-matrix membranes. *J. Memb. Sci.* **361**, 28–37 (2010).
45. William J. Koros, De Q. Vu, & Rajiv Mahajn, S. J. M. Mixed matrix membranes with pyrolized carbon sieve particles and methods of making and using the same. (2001).
46. Rezakazemi, M., Ebadi Amooghin, A., Montazer-Rahmati, M. M., Ismail, A. F. & Matsuura, T. State-of-the-art membrane-based CO_2 separation using mixed matrix membranes (MMMs): An overview on current status and future directions. *Prog. Polym. Sci.* **39**, 817–861 (2014).
47. Li, Y., Guan, H.-M., Chung, T.-S. & Kulprathipanja, S. Effects of novel silane modification of zeolite surface on polymer chain rigidification and partial pore blockage in polyethersulfone (PES)–zeolite A mixed matrix membranes. *J. Memb. Sci.* **275**, 17–28 (2006).
48. Husain, S. & Koros, W. J. Mixed matrix hollow fiber membranes made with modified HSSZ-13 zeolite in polyetherimide polymer matrix for gas separation. *J. Memb. Sci.* **288**, 195–207 (2007).
49. Li, Y., Chung, T.-S., Cao, C. & Kulprathipanja, S. The effects of polymer chain rigidification, zeolite pore size and pore blockage on polyethersulfone (PES)-zeolite A mixed matrix membranes. *J. Memb. Sci.* **260**, 45–55 (2005).

50. Zhan, X., Lu, J., Tan, T. & Li, J. Mixed matrix membranes with HF acid etched ZSM-5 for ethanol/water separation: Preparation and pervaporation performance. *Appl. Surf. Sci.* **259**, 547–556 (2012).
51. Aroon, M. A., Ismail, A. F., Matsuura, T. & Montazer-Rahmati, M. M. Performance studies of mixed matrix membranes for gas separation: A review. *Sep. Purif. Technol.* **75**, 229–242 (2010).
52. Cong, H., Radosz, M., Towler, B. F. & Shen, Y. Polymer–inorganic nanocomposite membranes for gas separation. *Sep. Purif. Technol.* **55**, 281–291 (2007).
53. Şen, D., Kalıpçılar, H. & Yilmaz, L. Development of polycarbonate based zeolite 4A filled mixed matrix gas separation membranes. *J. Memb. Sci.* **303**, 194–203 (2007).
54. Yong, H. H., Park, H. C., Kang, Y. S., Won, J. & Kim, W. N. Zeolite-filled polyimide membrane containing 2,4,6-triaminopyrimidine. *J. Memb. Sci.* **188**, 151–163 (2001).
55. Duval, J.-M. *et al.* Preparation of zeolite filled glassy polymer membranes. *J. Appl. Polym. Sci.* **54**, 409–418 (1994).
56. Mahajan, R. & Koros, W. J. Mixed matrix membrane materials with glassy polymers. Part 1. *Polym. Eng. Sci.* **42**, 1432–1441 (2002).
57. Venna, S. R. *et al.* Fabrication of MMMs with improved gas separation properties using externally-functionalized MOF particles. *J. Mater. Chem. A* **3**, 5014–5022 (2015).
58. Anjum, M. W. *et al.* Modulated UiO-66-based mixed-matrix membranes for CO_2 separation. *ACS Appl. Mater. Interfaces* **7**, 25193–201 (2015).
59. Smith, S. J. D. *et al.* Post-synthetic Ti exchanged UiO-66 metal-organic frameworks that deliver exceptional gas permeability in mixed matrix membranes. *Sci. Rep.* **5**, 7823 (2015).
60. Feijani, E. A., Tavasoli, A. & Mahdavi, H. Improving gas separation performance of poly(vinylidene fluoride) based mixed matrix membranes containing metal-organic frameworks by chemical modification (2015).
61. Naseri, M., Mousavi, S. F., Mohammadi, T. & Bakhtiari, O. Synthesis and gas transport performance of MIL-101/Matrimid mixed matrix membranes. *J. Ind. Eng. Chem.* **29**, 249–256 (2015).
62. Abedini, R. *et al.* Highly permeable poly(4-methyl-1-pentyne)/NH_2 -MIL 53 (Al) mixed matrix membrane for CO_2/CH_4 separation. *RSC Adv.* **4**, 36522 (2014).
63. Denny, M. S. & Cohen, S. M. In situ modification of metal-organic frameworks in mixed-matrix membranes. *Angew. Chemie Int. Ed.* **54**, 9029–9032 (2015).
64. Rodenas, T., van Dalen, M., Serra-Crespo, P., Kapteijn, F. & Gascon, J. Mixed matrix membranes based on NH_2-functionalized MIL-type MOFs: Influence of structural and operational parameters on the CO_2/CH_4 separation performance. *Microporous Mesoporous Mater.* **192**, 35–42 (2014).
65. Casado-Coterillo, C. *et al.* Synthesis and characterisation of MOF/ionic liquid/chitosan mixed matrix membranes for CO_2/N_2 separation. *RSC Adv.* **5**, 102350–102361 (2015).
66. Noble, R. D. Perspectives on mixed matrix membranes. *Journal of Membrane Science* **378**, 393–397 (2011).
67. Hudiono, Y. C. *et al.* A three-component mixed-matrix membrane with enhanced CO_2 separation properties based on zeolites and ionic liquid materials. *J. Memb. Sci.* **350**, 117–123 (2010).

68. Mohshim, D. F., Mukhtar, H. & Man, Z. The effect of incorporating ionic liquid into polyethersulfone-SAPO34 based mixed matrix membrane on CO_2 gas separation performance. *Sep. Purif. Technol.* **135**, 252–258 (2014).
69. Nasir, R., Mukhtar, H., Man, Z. & Mohshim, D. F. Material Advancements in Fabrication of Mixed-Matrix Membranes. *Chem. Eng. Technol.* **36**, 717–727 (2013).
70. Gorgojo, P., Uriel, S., Téllez, C. & Coronas, J. Development of mixed matrix membranes based on zeolite Nu-6(2) for gas separation. *Microporous Mesoporous Mater.* **115**, 85–92 (2008).
71. Xu, L. Interfacial engineering of the interphase between carbon fibers and vinyl ester resin. (2003).
72. Broyles, N. S. *et al.* Fatigue performance of carbon fibre/vinyl ester composites: the effect of two dissimilar polymeric sizing agents. *Polymer (Guildf).* **39**, 3417–3424 (1998).
73. Marti, A.M., Venna, S.R., Roth, E.A., Culp, J.T. & Hopkinson, D.P., Simple fabrication method for mixed matrix membranes with in situ MOF growth for Gas separation. *ACS Applied Materials and Interfaces* **10**, 24784–24790 (2018).
74. Kessler, J. O. & Moody, C. D. Drinking water from sea water by forward osmosis. *Desalination* **18**, 297–306 (1976).
75. Elimelech, M. Yale constructs forward osmosis desalination pilot plant. *Membr. Technol.* **2007**, 7–8 (2007).
76. Achilli, A., Cath, T. Y., Marchand, E. A. & Childress, A. E. The forward osmosis membrane bioreactor: A low fouling alternative to MBR processes. *Desalination* **239**, 10–21 (2009).
77. Cornelissen, E. R. *et al.* Membrane fouling and process performance of forward osmosis membranes on activated sludge. *J. Memb. Sci.* **319**, 158–168 (2008).
78. Holloway, R. W., Childress, A. E., Dennett, K. E. & Cath, T. Y. Forward osmosis for concentration of anaerobic digester centrate. *Water Res.* **41**, 4005–4014 (2007).
79. Basu, S., Cano-Odena, A. & Vankelecom, I. F. J. Asymmetric Matrimid®/ [$Cu_3(BTC)_2$] mixed-matrix membranes for gas separations. *J. Memb. Sci.* **362**, 478–487 (2010).
80. Junaidi, M. U. M., Leo, C. P., Ahmad, A. L. & Ahmad, N. A. Fluorocarbon functionalized SAPO-34 zeolite incorporated in asymmetric mixed matrix membranes for carbon dioxide separation in wet gases. *Microporous Mesoporous Mater.* **206**, 23–33 (2015).
81. Peng, N., Widjojo, N., Sukitpaneenit, P., Teoh, M.M., Lipscomb, G.G., Chung, T.-S. & Lai, J.-Y., Evolution of polymeric hollow fibers as sustainable technologies: Past, present, and future. *Progress in Polymer Science* **37**, 1401–1424 (2012).
82. Dai, Y., Johnson, J. R., Karvan, O., Sholl, D. S. & Koros, W. J. Ultem®/ZIF-8 mixed matrix hollow fiber membranes for CO_2/N_2 separations. *J. Memb. Sci.* **401**, 76–82 (2012).
83. Wang, Z. & Cohen, S. M. Postsynthetic modification of metal-organic frameworks. *Chem. Soc. Rev.* **38**, 1315–29 (2009).
84. Adams, R., Carson, C., Ward, J., Tannenbaum, R. & Koros, W. Metal organic framework mixed matrix membranes for gas separations. *Microporous Mesoporous Mater.* **131**, 13–20 (2010).

85. Xin, Q. *et al.* Mixed matrix membranes composed of sulfonated poly(ether ether ketone) and a sulfonated metal-organic framework for gas separation. *J. Memb. Sci.* **488**, 67–78 (2015).
86. Venna, S.R., Spore, A., Marti, A.M., Albenze, E.J., Nulwala, H.B., Rosi, N.L., Luebke, D.R., Hopkinson, D.P. & Allcock, H.R., Polyphosphazene polymer development for mixed matrix membranes using SIFSIX-Cu-2*i* as performance enhancement filler particles, *Journal of Membranes Science* **535**, 103–112 (2017).
87. Waqas Anjum, M., Bueken, B., De Vos, D. & Vankelecom, I. F. J. MIL-125(Ti) based mixed matrix membranes for CO_2 separation from CH_4 and N_2. *J. Memb. Sci.* **502**, 21–28 (2016).
88. Guo, X. *et al.* Mixed matrix membranes incorporated with amine-functionalized titanium-based metal-organic framework for CO_2/CH_4 separation. *J. Memb. Sci.* **478**, 130–139 (2015).
89. Hsieh, J. O., Balkus, K. J., Ferraris, J. P. & Musselman, I. H. MIL-53 frameworks in mixed-matrix membranes. *Microporous Mesoporous Mater.* **196**, 165–174 (2014).
90. H. Yehia, T. J. Pisklak, J. P. Ferraris, K. J. Balkus, I. H. M. Methane facilitated transport using copper (II) biphenyl dicarboxylate-triethylenediamine poly(3-acetoxyethylthiophene) mixed matrix membranes. *Polym. Prepr.* **45**, 35–36 (2004).
91. Perez, E., Karunaweera, C., Musselman, I., Balkus, K. & Ferraris, J. Origins and evolution of inorganic-based and MOF-based mixed-matrix membranes for gas separations. *Processes* **4**, 32 (2016).
92. Car, A., Stropnik, C. & Peinemann, K.-V. Hybrid membrane materials with different metal-organic frameworks (MOFs) for gas separation. *Desalination* **200**, 424–426 (2006).
93. Díaz, K., Garrido, L., López-González, M., Castillo, L. F. del & Riande, E. CO_2 transport in polysulfone membranes containing zeolitic imidazolate frameworks as determined by permeation and PFG NMR techniques (2009).
94. Christian Serre *et al.* Very large breathing effect in the first nanoporous chromium(III)-based solids: MIL-53 or CrIII(OH) · $\{O_2C–C_6H_4–CO_2\}$ · $\{HO_2C–C_6H_4–CO_2H\}$ x · H_2Oy (2002).
95. Basu, S., Cano-Odena, A. & Vankelecom, I. F. J. MOF-containing mixed-matrix membranes for CO_2/CH_4 and CO_2/N_2 binary gas mixture separations. *Sep. Purif. Technol.* **81**, 31–40 (2011).
96. Zornoza, B. *et al.* Functionalized flexible MOFs as fillers in mixed matrix membranes for highly selective separation of CO_2 from CH_4 at elevated pressures. *Chem. Commun.* **47**, 9522 (2011).
97. Thompson, J. A. *et al.* Mixed-linker zeolitic imidazolate framework mixed-matrix membranes for aggressive CO_2 separation from natural gas. *Microporous Mesoporous Mater.* **192**, 43–51 (2014).
98. Erucar, I., Yilmaz, G. & Keskin, S. Recent advances in metal-organic framework-based mixed matrix membranes. *Chem. Asian J.* **8**, 1692–704 (2013).
99. Bae, T.-H. *et al.* A high-performance gas-separation membrane containing submicrometer-sized metal-organic framework crystals. *Angew. Chem. Int. Ed. Engl.* **49**, 9863–6 (2010).
100. Nafisi, V. & Hägg, M.-B. Development of dual layer of ZIF-8/PEBAX-2533 mixed matrix membrane for CO_2 capture. *J. Memb. Sci.* **459**, 244–255 (2014).

101. Perez, E. V., Balkus, K. J., Ferraris, J. P. & Musselman, I. H. Mixed-matrix membranes containing MOF-5 for gas separations. *J. Memb. Sci.* **328**, 165–173 (2009).
102. Yang, T., Xiao, Y. & Chung, T.-S. Poly-/metal-benzimidazole nano-composite membranes for hydrogen purification. *Energy Environ. Sci.* **4**, 4171 (2011).
103. Rodenas, T. *et al.* Visualizing MOF mixed matrix membranes at the nanoscale: Towards structure-performance relationships in CO_2/CH_4 separation over NH_2-MIL-53(Al)@PI. *Adv. Funct. Mater.* **24**, 249–256 (2014).
104. Zornoza, B. *et al.* Functionalized flexible MOFs as fillers in mixed matrix membranes for highly selective separation of CO_2 from CH_4 at elevated pressures. *Chem. Commun. (Camb).* **47**, 9522–4 (2011).
105. Burmann, P., Zornoza, B., Téllez, C. & Coronas, J. Mixed matrix membranes comprising MOFs and porous silicate fillers prepared via spin coating for gas separation. *Chem. Eng. Sci.* **107**, 66–75 (2014).
106. Japip, S., Wang, H., Xiao, Y. & Shung Chung, T. Highly permeable zeolitic imidazolate framework (ZIF)-71 nano-particles enhanced polyimide membranes for gas separation. *J. Memb. Sci.* **467**, 162–174 (2014).
107. Zhang, C. *et al.* Highly scalable ZIF-based mixed-matrix hollow fiber membranes for advanced hydrocarbon separations. *AIChE J.* **60**, 2625–2635 (2014).
108. Peydayesh, M., Asarehpour, S., Mohammadi, T. & Bakhtiari, O. Preparation and characterization of SAPO-34 — Matrimid® 5218 mixed matrix membranes for CO_2/CH_4 separation. *Chem. Eng. Res. Des.* **91**, 1335–1342 (2013).
109. De Q Vu, William J Koros, S. J. M. Mixed matrix membranes using carbon molecular sieves: II. Modeling permeation behavior. *J. Memb. Sci.* **211**, 335–348 (2003).
110. Zhao, Z., Ma, X., Li, Z. & Lin, Y. S. Synthesis, characterization and gas transport properties of MOF-5 membranes. *J. Memb. Sci.* **382**, 82–90 (2011).
111. Perez, E. V., Balkus, K. J., Ferraris, J. P. & Musselman, I. H. Mixed-matrix membranes containing MOF-5 for gas separations. *J. Memb. Sci.* **328**, 165–173 (2009).
112. Seoane, B. *et al.* Metal-organic framework based mixed matrix membranes: A solution for highly efficient CO_2 capture? *Chem. Soc. Rev.* **44**, 2421–2454 (2015).
113. Zornoza, B. *et al.* Metal organic framework based mixed matrix membranes: An increasingly important field of research with a large application potential. *Microporous Mesoporous Mater.* **166**, 67–78 (2013).
114. Xin, Q. *et al.* Enhanced interfacial interaction and CO_2 separation performance of mixed matrix membrane by incorporating polyethylenimine-decorated metal-organic frameworks. *ACS Appl. Mater. Interfaces* **7**, 1065–1077 (2015).
115. Pal, R. Permeation models for mixed matrix membranes. *J. Colloid Interface Sci.* **317**, 191–8 (2008).
116. Watanabe, T. & Sholl, D. S. Accelerating applications of metal-organic frameworks for gas adsorption and separation by computational screening of materials. *Langmuir* **28**, 14114–28 (2012).
117. Bouma, R. H. B., Checchetti, A., Chidichimo, G. & Drioli, E. Permeation through a heterogeneous membrane: the effect of the dispersed phase. *J. Memb. Sci.* **128**, 141–149 (1997).
118. Maxwell, J. C. *Treatise on Electricity and Magnetism.* (Oxford University Press, London, 1873).

119. Petropoulos, J. H. A comparative study of approaches applied to the permeability of binary composite polymeric materials. *J. Polym. Sci. Polym. Phys. Ed.* **23**, 1309–1324 (1985).
120. Bruggeman, D. A. G. Berechnung verschiedener physikalischer Konstanten von heterogenen Substanzen. I. Dielektrizitätskonstanten und Leitfähigkeiten der Mischkörper aus isotropen Substanzen. *Ann. Phys.* **416**, 636–664 (1935).
121. Nielsen, L. E. Thermal conductivity of particulate-filled polymers. *J. Appl. Polym. Sci.* **17**, 3819–3820 (1973).
122. Lewis, T. B. & Nielsen, L. E. Dynamic mechanical properties of particulate-filled composites. *J. Appl. Polym. Sci.* **14**, 1449–1471 (1970).
123. Pal, R. New models for thermal conductivity of particulate composites. *J. Reinf. Plast. Compos.* **26**, 643–651 (2007).
124. Felske, J. D. Effective thermal conductivity of composite spheres in a continuous medium with contact resistance. *Int. J. Heat Mass Transf.* **47**, 3453–3461 (2004).
125. Shimekit, B., Mukhtar, H. & Murugesan, T. Prediction of the relative permeability of gases in mixed matrix membranes. *J. Memb. Sci.* **373**, 152–159 (2011).
126. Theodorou, D. N. & Suter, U. W. Detailed molecular structure of a vinyl polymer glass. *Macromolecules* **18**, 1467–1478 (1985).
127. Qiang Wang, Qiliang Yan, Paul F. Nealey, & Pablo*, J. J. de. Monte Carlo simulations of diblock copolymer thin films confined between chemically heterogeneous hard surfaces (2000).
128. Feng, J. & Ruckenstein, E. The morphology of symmetric triblock copolymer melts confined in a slit: A Monte Carlo simulation. *Macromol. Theory Simulations* **11**, 630 (2002).
129. Yu, B. *et al.* Confinement-induced novel morphologies of block copolymers. *Phys. Rev. Lett.* **96**, 138306 (2006).
130. Feng, J. & Ruckenstein*, E. Morphologies of AB diblock copolymer melts confined in nanocylindrical tubes (2006).
131. Feng, J. & Ruckenstein, E. Morphology transitions of AB diblock copolymer melts confined in nanocylindrical tubes. *J. Chem. Phys.* **125**, 164911 (2006).
132. Feng, J., Cavicchi, K. A. & Heinz, H. Control over self-assembly of diblock copolymers on hexagonal and square templates for high area density circuit boards. *ACS Nano* **5**, 9413–20 (2011).
133. Termonia, Y. Monte-Carlo modeling of dense polymer melts near nanoparticles. *Polymer (Guildf).* **50**, 1062–1066 (2009).
134. Gao, Y., Liu, J., Zhang, L. & Cao, D. Existence of a glassy layer in the polymer-nanosheet interface: Evidence from molecular dynamics. *Macromol. Theory Simulations* **23**, 36–48 (2014).
135. Feng, J., Venna, S. R. & Hopkinson, D. P. Interactions at the interface of polymer matrix-filler particle composites. *Polymer (Guildf).* **103**, 189–195 (2016).
136. Budhathoki, S., Ajayi, O., Steckel, J.A. & Wilmer, C.E., High-throughput computational prediction of the cost of carbon capture using mixed matrix membranes, *Energy Environmental Science* **12**, 1255–1264 (2019).

Chapter 10

MOF-Stability, Challenges and Outlook

10.1. Metal Organic Frameworks stability

10.1.1. *Generalities*

Many MOFs have shown great promise for a wide variety of applications, including MOF membranes for molecular gas separations, but ultimately its potential industrial applications may be limited because of stability issues. Under typical industrial-like conditions (i.e. presence of impurities, presence of water/moisture, high pressure, high temperature, etc) MOFs must retain their framework integrity to maintain their intended functionality. The stability of MOFs can be affected by several factors, including the operating conditions, metal ions, organic linkers, metal–ligand coordination geometry, pore surface hydrophobicity, among others.[1] Comprehensive reviews on MOF stability have been documented.[1–3] In the next sections. we summarize the main different strategies that have been employed to improve MOF stability in general, and then, we focus specifically on how some of these approaches, and other novel approaches have been extended to enhance MOF membrane stability.

10.1.2. *Strategies to improve MOF stability*

In general, a stable MOF structure should possess strong coordination to resist the attack of guest molecules or possess steric hindrance to prevent the intrusion of guest molecules to the metal nodes.[1] Several strategies able to improve the stability of MOFs have been reviewed recently.[2] These strategies consider the: (a) strength of metal ligand bonds, (b) chemical functionality of linkers, (c) self-strengthening of the framework, and (d) surface and post synthesis modification.

(a) *Metal ligand bonds*

The metal–ligand bond strength determines the thermodynamic stability of MOFs under many operating environments.[1] In this respect, the stability of MOFs can be predicted by examining the strength of the bonds that form the framework. In general, a more stable MOF framework is formed when high-valent metal ions such as Ti^{4+}, Zr^{4+}, Al^{3+}, Fe^{3+}, and Cr^{3+} with high charge densities are used, since they can form stronger coordination bonds. In particular, these metal ions can form stable MOFs with carboxylate-based ligands (hard bases). Stable MOFs can also be formed by coordinating soft divalent metal ions such as Zn^{2+}, Cu^{2+}, Ni^{2+}, Mn^{2+}, and Ag+ with soft azolate ligands such as imidazolates, pyrazolates, triazolates, and tetrazolates. The most representative family of these stable MOFs are the so-called zeolitic imidazolate frameworks. Therefore, to form MOFs with enhanced water stability, the coordination of carboxylate-based ligands (hard Lewis bases) and high-valent metal ions (hard Lewis acids), or azolate-based ligands (soft Lewis bases) and low-valence transition metal ions (soft Lewis acids) is highly preferred. This helps to enhance the strength of the coordination bonds between inorganic clusters and coordination groups.

As compared to neutral water molecules, acid and bases are more destructive to MOFs. In general, MOFs with high-valence metal ions and carboxylate-based ligands display enhanced stability in the presence of acid solutions;[1] while low-valence metal ions coordinated with azolate groups are more stable in basic solutions due to their low affinity to OH^- groups.[4,5]

(b) *Linker chemical functionality*

Several studies have demonstrated that the stability of MOFs can be improved by modifying the chemical functionality of the linkers, allowing their effective use in applications in which water/moisture is present. For instance, the incorporation of hydrophobic functional groups including alkyl or fluorinated groups into the MOF channels can enhance their hydrophobic properties, and consequently improve their water stability. Specifically, the inclusion of methyl groups,[6,7] trifluoromethoxy functional groups,[8] 3,5-bis(trifluoromethyl)-1,2,4triazolate linkers,[9] hydrophobic/nonpolar functional groups,[10] fluorine[11] in the MOF ligand have helped to enhanced the hydrophobicity of the MOF and therefore its water stability.

It is important to mention that although the strategy of chemically modifying the functionality of the MOF linkers can improve the MOF stability, a clear disadvantage of this approach is the fact that textural properties such as surface area, and pore volume, as well as structural integrity of the resultant MOF, may be compromised.

(c) *Framework self-strengthening*

The formation of multi-walled MOFs is another alternative approach which has been employed to improve their stability. Due to the increase in wall thickness, the multi-walled MOF architecture typically results in a high degree of structural robustness and mechanical stability.[2] Some representative examples illustrating the synthesis of multi-walled MOFs have been reported elsewhere.[12–15] The synthesis of interpenetrated frameworks is another approach which has been demonstrated to improve mainly thermal stability of MOFs. In general, this strategy increases the wall thickness, and reduces the pore size of MOFs. Both of these features lead to a thermal stability improvement. This concept has been exploited for the synthesis of several MOF compositions.[16–19] The use of preformed metal–oxo clusters with structures identical to the secondary building units of the targeted MOF can help to improve its stability. MIL type MOFs,[20] Zr-based MOFs,[21] iron, cobalt, nickel, manganese, and zinc-based MOFs[22] have been successfully synthesized using preformed clusters.

(d) *Surface and post-synthesis modification*

MOF surface modification can afford not only the preparation of MOFs with desired functionalities, but also an effective way to improve stability. For instance, post-synthetic ligand and metal ion exchange have been demonstrated to be effective approaches to prepare MOFs with enhanced stability. Post-synthetic ligand and metal ion exchange has been demonstrated for ZIFs, MIL series, and UiO series.[23] Some examples of using surface modification to improve the overall stability of MOFs include the use of plasma enhanced chemical deposition,[24] polydimethylsiloxane,[25] and graphene oxide to hybridized MOFs.[26]

10.1.3. *Representative membrane MOF stability studies*

MOF membrane stability has been identified as one of the major challenges in the MOF membrane community. As pointed out by Professor Lin.[27] "Understanding the thermal and chemical stability of the MOF membrane materials in the gas phases, and if unstable, develop methods to improve the stability of the materials, will be critical to advancing MOF membrane technology for gas separation". Few studies focused on the chemical, thermal, and long term stability of MOF membranes for gas separations have been documented. These studies mainly focus on two types of MOF membranes: MOF 5 [28] and ZIF-8[29–33] and ZIF-90.[34] In the next few paragraphs, we summarize the central results of these studies.

MOF-5 membranes were evaluated for the pervaporation and separation of toluene (T) , o-xylene (X) and 1,3,5 triisopropylbenzene (TIPB).[28] The membranes displayed a continuous decrease of permeation flux and separation factor with stream time. Authors attributed this decrease to the presence of immobile permeating molecules within the MOF-5 structure. Nevertheless, MOF-5 membranes were structurally stable in the presence of T, X, and TIPB. Authors recognized the need of addressing the observed fouling issue for pervaporation separation of organic liquids.

As compared to MOF-5 membranes, ZIF membranes display enhanced chemical and thermal which is mainly associated to the strong bonding between center metal ions (Zn or Co) and nitrogen atoms in ligands. Lai's group demonstrated the thermal stability of α-alumina supported ZIF-8 membranes.[29] Specifically, these authors heated the ZIF-8 membranes to $\sim$180°C and then cooled down to 15°C; and evaluated their separation performance for propane/propylene mixtures. The authors found only a slight change in separation performance upon this thermal gradient ($>$2%), suggesting good thermal stability of the ZIF-8 membranes. Lin's group reported the long term stability for propane/propylene mixtures.[30] In this study, the authors found that the permeance and separation selectivity of the membranes remained constant for over 35 days of off-stream storage in laboratory atmosphere and 5 days on stream testing. These results indicate the structural stability of the ZIF-8 membranes under atmospheric moisture.

Lin's group studied the water stability of ZIF-8 membranes through pervaporation and batch durability tests at room temperature.[31] The authors found that ZIF-8 membranes underwent degradation during water pervaporation. This observation was confirmed experimentally by the sharp increase in water pervaporation flux and the disappearance of ZIF-8 structure after pervaporation. To have a better understanding on the mechanism of water stability of ZIF-8 structure, the authors carried out hydrolysis control experiments on ZIF-8. Authors concluded that ZIF-8 crystals degrade in water due to the dissolution of ZIF-8. The rate and extent of this dissolution depended on the ZIF-8 to water mass ratio.

To alleviate the limited water stability of ZIF-8 membranes, the same group demonstrated that the stability of ZIF-8 membranes can be greatly improved by incorporating a hydrophobic organic linker in the ZIF-8 structure via surface ligand exchange.[32] Specifically, in this approach authors employed a ligand exchange post-modification method that replaced methylimidazole ligands (organic linker of ZIF-8) on the outer surface of ZIF-8 membranes by a more hydrophobic, bulkier 5,6-dimethylbenzimidazole ligand. Interestingly, this ligand exchange did not change the crystal structure, morphology or gas permeance of the ZIF-8 membranes. This ligand exchange method is a promising approach that resulted in hydrostable ZIF-8 membranes with enhanced surface hydrophobicity.

The thermal stability of ZIF-8 membranes at high temperature has been documented.[33] In this paper, authors carried out dynamic temperature dependent permeation experiments for H_2/CO_2 mixtures. Authors found that H_2/CO_2 permeance and separation selectivities followed the solution-diffusion model at temperatures in the 25 to 275°C range. At higher temperatures, authors suggested the formation of an amorphous carbonized imidazole-Zn-azirine structure. In general, ZIF-8 membranes preserved their crystalline integrity and separation performance at temperatures below 150°C. Above this temperature, ZIF-8 membranes experienced carbonization of the imidazole ligand. This carbonization process modified the pore structure of the ZIF-8 membrane, and therefore its separation performance for H_2/CO_2.

Caro's group reported thermal and long term stability studies for ZIF-90 membranes.[34] Specifically, this group demonstrated stable hydrogen permeance and H_2/CH_4 separation selectivity at 200°C during 24 hours. These studies show that ZIF membranes display moderate to high structural, chemical, and thermal stability.

10.2. Challenges and outlook

A decade has passed since the first report of the successful synthesis of a continuous metal organic framework membrane demonstrating ability to separate binary gas mixtures. Since then, considerable research progress has been done in the development of metal-organic framework membranes for molecular gas separations. Some of these membranes display separation performances which are appealing for its potential use at industrial level for diverse relevant and challenging molecular gas separations. Most of these membranes display overall good separation performance for *binary* gas mixtures. However, practical challenges that may prevent its potential industrial application need to be addressed. Key aspects that require attention from the scientific community to envision these membranes as an economic viable technology to effectively separate gas mixtures are briefly highlighted in the next few paragraphs.

(a) *Realistic separation conditions*

Realistic separation conditions for any industrially relevant gas separation require the membrane to perform well at high pressures and or temperatures and in the presence of other gas impurities. In principle, the overall separation performance of MOF membranes under high pressure and/or temperature as well as in the presence of gas impurities, should decrease. To our best knowledge, there are no reports in the open literature addressing the stability and performance of MOF membranes under real gas processing conditions (presence of impurities, high temperature, and pressure).

(b) *Membrane reproducibility*

Limited reported data on membrane reproducibility is a common problem found in MOF membranes. Current and future studies on MOF membranes should address this issue. The separation index (π)[35,36] which considers both the separation selectivity and gas permeance has been used as a reliable metric to estimate/quantify membrane reproducibility for some ZIF-8 membranes.[37,38] Other metrics would be useful to assess membrane reproducibility.

(c) *Quantifying and understanding membrane defects*

The presence of non-selective pathways and/or membrane defects play a critical role in the overall separation performance of the membranes. A fundamental understanding on how these defects affect membrane performance may help in rationally designing membranes with enhanced separation performance. There is a critical need to find reliable methods to estimate and quantify the fraction of gas flux through defects (intercrystalline pathways larger than the "selective pores in MOF membranes). One may think that the methods that have been successfully employed to characterize defects in zeolite membranes may apply for MOF membranes too. These methods include mainly permporosimetry, fluorescence confocal optical microscopy, measuring the flux of a molecule too large to fit in the zeolite pores, measuring the ratio of single gas permeances, and temperature programmed desorption of alkanes. The most common method to quantify flow through defects is permporosimetry,[39–41] in which the flow of helium is measured as a function of a condensable hydrocarbon which should have size around the pore size of the zeolite. In addition, this hydrocarbon should adsorb in the micropore strongly so that helium cannot enter the porous structure. Hydrocarbon adsorption blocks selective pores and the remaining flux is assumed to be through defects. Temperature programmed desorption (TPD) of alkanes has been used successfully to quantify directly the volume of non-zeolite pores.[42]

(d) *Membrane scale-up studies*

Once promising membrane compositions displaying good separation performance for any gas mixture have been identified at laboratory level, scale-up studies may be required to assess the potential use of these membranes at pilot plant scale. Scaling-up membrane synthesis is not a trivial issue, and requires considerable effort from the scientific community. Membrane scale-up studies are needed to assess the potential effectiveness of membranes for molecular gas separations not only at laboratory scale, but also at pilot plant level.

(e) *Cost*

The relatively high cost associated with the membrane support may limit its potential practical gas separations. Typically, the support heavily accounts for the higher total membrane cost. Therefore, research efforts should consider the use of cheaper membrane supports. In addition, the economics of any separation technology needs to be evaluated in the context of metrics such as cost per kilogram of product and energy use per kilogram.[43]

(f) *Novel MOF membrane compositions*

Developing continuous novel MOF membrane compositions with suitable pore sizes (between the sizes of the molecules to be separated) should lead to "sharper" molecular sieving effect, and therefore to higher separation selectivities. In addition, these membrane compositions should be stable in the presence of water and other gases typically encountered in several relevant gas mixtures, including hydrocarbons, sulfur, etc. "Exploiting the main separation mechanisms (molecular sieving, differences in diffusivities or kinetic contribution, and competitive adsorption or thermodynamic contribution) proposed for microporous crystalline membranes,[44] should lead to highly selective and highly permeable MOF membranes for numerous industrially relevant gas separations". Figure 10.1 summarizes the typical governing separation mechanisms of gas mixtures over microporous crystalline sieve membranes.

(g) *Functionalized MOF membranes*

The adsorption capacity of different gases over MOF membranes can be fine-tuned via surface modification. Ideally, the preferential adsorption of the gas to be separated should be promoted. In principle, this would result in higher diffusivity of the "selective gas". Therefore, if the preferential *selective gas adsorption selectivity is enhanced*, higher separation selectivities can be obtained. Modifying the organic linker functionality may help to enhance this selectivity gas preferential adsorption.

Mixed matrix membranes possess great potential to significantly improve the gas transport properties compared to their parent polymer matrix without much increase in the cost of processing. MMM research gained momentum with MOFs. MOF-based MMM have been used to

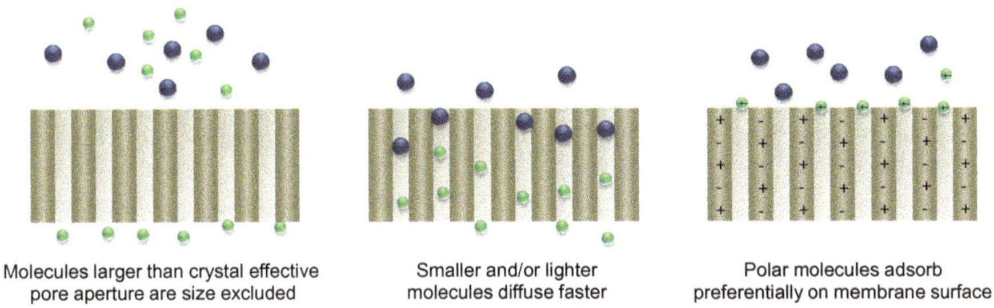

(a) Molecular Sieving (b) Diffusivity differences (Kinetic contribution) (c) Competitive Adsorption (Thermodynamic contribution)

Figure 10.1. Separation mechanisms of gas mixtures over porous crystalline sieve membranes: (a) Molecular sieving, (b) Diffusivity differences, and (c) Competitive adsorption. Green spheres represent the smallest and most polar molecule to be recovered in the permeate side. For simplicity a binary mixture is presented. (Modified from reference 44).

separate several gases such as CO_2, N_2, CH_4, H_2, O_2, He, H_2O, olefins, parafins, etc. The selection of polymers and type of MOF plays a major role in the fabrication of MMM and dictate the overall properties of MMM for specific gas pair separation. Research has focused mainly on polymers with low gas permeability. Current research efforts must focus on the polymers that are close to Robeson upper bound to obtain the industrially viable gas separation performance with the MMM. However, successful industrial implementation of these mixed matrix membranes for commercial separations has not yet been achieved because of several problems related to their processability into thin films on the porous supports. The challenging factor for the fabrication and understanding of these membranes includes the synthesis of smaller particles, compatibilize the polymer-particle, characterization of the interface, etc. Compared to zeolite, tunable MOF chemistry allowed to synthesize nano MOFs, which are very useful to fabricate thin membranes. Most importantly, MMM should be studied at industrial conditions in order to analyze their stability to routinely apply MMM in the industry and focus the research in improving their stability.

If these challenges and issues are properly tackled, one may see in the near future some of these MOF membranes as a feasible economic and effective technology employed to separate challenging gas mixtures at industrial relevant conditions.

References

1. Yuan, S., Feng, L., Wang, K., Pang, J., Bosch, M., Lollar, C., Sun, Y., Qin, J., Yang, X., Zhang, P., Wang, Q., Zou, L., Zhang, Y., Zhang, L., Fang, Y., Li, J. & Zhou, H. C. Stable metal-organic frameworks: Design, synthesis, and applications. *Adv. Mater.* **30**, 1704303 (2018).
2. Li, N., Xu, J., Feng, R., Hu, T. L. & Bu, X. H. Governing metal-organic frameworks towards high stability. *Chem. Commun.* **52**, 8501–8513 (2016).
3. Canivet, J., Fateeva, A., Guo, Y., Coasne, B. & Farrusseng, D. Water adsorption in MOFs: fundamentals and applications. *Chem. Soc. Rev.* **43**, 5594 (2014).
4. Furukawa, H., Gándara, F., Zhang, Y.-B., Jiang, J., Queen, W. L., Hudson, M. R. & Yaghi, O. M. Water adsorption in porous metal-organic frameworks and related materials. *J. Am. Chem. Soc.* **136**, 4369 (2014).
5. Wang, K., Lv, X. L., Feng, D., Li, J., Chen, S., Sun, J., Song, L., Xie, Y., Li, J. R. & Zhou, H. C. Pyrazolate-based porphyrinic metal-organic framework with extraordinary base-resistance. *J. Am. Chem. Soc.* **138**, 914 (2016).
6. Ma, D., Li, Y. & Li, Z. Tuning the moisture stability of metal–organic frameworks by incorporating hydrophobic functional groups at different positions of ligands. *Chem. Commun.* **47**, 7377 (2011).
7. Yang, J., Grzech, A., Mulder, F. M. & Dingemans, T. J. Methyl Modified MOF-5: A Water Stable Hydrogen Storage Material. *Chem. Commun.* **47**, 5244 (2011).
8. Wu, T., Shen, L., Luebbers, M., Hu, C., Chen, Q., Ni, Z. & Masel, R. I. Enhancing the stability of metal–organic frameworks in humid air by incorporating water repellent functional groups. *Chem. Commun.* **46**, 6120 (2010).
9. Yang, C., Wang, X. & Omary, M. A. Fluorous Metal–Organic Frameworks for High-Density Gas Adsorption. *J. Am. Chem. Soc.* **129**, 15454 (2007).
10. Jasuja, H., Huang, Y. G. & Walton, K. S. Adjusting the Stability of Metal-Organic Frameworks Under Humid Conditions by Ligand Functionalization. *Langmuir* **28**, 16874 (2012).
11. Zhang, D. S., Chang, Z., Li, Y. F., Jiang, Z. Y., Xuan, Z. H., Zhang, Y. H., Li, J. R., Chen, Q., Hu, T. L. & Bu, X. H. Fluorous Metal-Organic Frameworks With Enhanced Stability and High H_2/CO_2 Storage Capacities. *Sci. Rep.* **3**, 3312 (2013).
12. Choi, H. J., Dinca, M., Dailly, A. & Long, J. R. Hydrogenstorage in water-stable metal–organic frameworks incorporating 1,3- and 1,4-benzenedipyrazolate. *Energy Environ. Sci.* **3**, 117 (2010).
13. Tian, D., Chen, Q., Li, Y., Zhang, Y. H., Chang, Z. & Bu, X. H. A Mixed Molecular Building Block Strategy for the Design of Nested Polyhedron Metal-Organic Frameworks. *Angew. Chem., Int. Ed.* **53**, 837 (2014).
14. Chen, K. J., Perry IV, J. J., Scott, H. S., Yang, Q. Y. & Zaworotko, M. J. Double-walled pyr Topology Networks From a Novel Fluoride-Bridged Heptanuclear Metal Cluster. *Chem. Sci.* **6**, 4784 (2015).
15. Tian, D., Xu, J., Xie, Z. J., Yao, Z. Q., Fu, D. L., Zhou, Z. & Bu, X. H. The First Example of Hetero-Triple-Walled Metal-Organic Frameworks With High Chemical Stability Constructed via Flexible Integration of Mixed Molecular Building Blocks. *Adv. Sci.* 1500283 (2015).

16. Dinca, M., Dailly, A., Tsayand, C. & Long, J. R. Expanded Sodalite-Type Metal-Organic Frameworks: Increased Stability and H_2 Adsorption Through Ligand-Directed Catenation. *Inorg.Chem.* **47**, 11 (2008).
17. Gao, W. Y., Zhang, Z., Cash, L., Wojtas, L., Chen, Y. S. & Ma, S. Two rare indium-based porous metal — metalloporphyrin frameworks exhibiting interesting CO_2 uptake. *CrystEngComm* **15**, 9320 (2013).
18. Rowsell, J. L. C. & Yaghi, O. M. Effects of Functionalization, Catenation, and Variation of the Metal Oxide and Organic Linking Units on the Low-Pressure Hydrogen Adsorption Properties of Metal-Organic Frameworks. *J. Am. Chem. Soc.* **128**, 1304 (2006).
19. Ma, S., Wang, X. S., Yuan, D. & Zhou, H. C. A Coordinatively Linked Yb Metal-Organic Framework Demonstrates High Thermal Stability and Uncommon Gas-Adsorption Selectivity. *Angew. Chem., Int. Ed.* **47**, 4130 (2008).
20. Devic, T. & Serre, C. High valence 3p and transition metal based MOFs. *Chem. Soc. Rev.* **43**, 6097 (2014).
21. Guillerm, V., Gross, S., Serre, C., Devic, T., Bauer, M. & Férey, G. A zirconium methacrylate oxocluster as precursor for the low-temperature synthesis of porous zirconium(iv) dicarboxylates. *Chem. Commun.* **46**, 767 (2010).
22. Feng, D., Wang, K., Wei, Z., Chen, Y. P., Simon, C. M., Arvapally, R. K., Martin, R. L., Bosch, M., Liu, T. F., Fordham, S., Yuan, D., Omary, M. A., Haranczyk, M., Smit, B. & Zhou, H. C. Kinetically Tuned Dimensional Augmentation as a Versatile Synthetic Route Towards Robust Metal-Organic Frameworks. *Nat. Commun.* **5**, 5723 (2014).
23. Kim, M., Cahill, J. F., Fei, H., Prather, K. A. & Cohen, S. M. Postsynthetic Ligand and Cation Exchange in Robust Metal-Organic Frameworks. *J. Am. Chem. Soc.* **134**, 18082 (2012).
24. Decoste, J. B., Peterson, G. W., Smith, M. W., Stone, C. A. & Willis, C. R. Enhanced Stability of Cu-BTC MOF via Perfluorohexane Plasma-Enhanced Chemical Vapor Deposition. *J. Am. Chem. Soc.* **134**, 1486 (2012).
25. Zhang, W., Hu, Y., Ge, J., Jiang, H. L. & Yu, S. H. A Facile and General Coating Approach to Moisture/Water-Resistant Metal-Organic Frameworks With Intact Porosity. *J. Am. Chem. Soc.* **136**, 16978 (2014).
26. Shen, Y., Li, Z., Wang, L., Ye, Y., Liu, Q., Ma, X., Chen, Q., Zhang, Z. & Xiang, S. Cobalt–citrate framework armored with graphene oxide exhibiting improved thermal stability and selectivity for biogas decarburization. *J. Mater. Chem. A*, **3**, 593 (2015).
27. Lin, Y. S. Metal organic framework membranes for separation applications. *Current Opinion in Chemical Engineering* **8**, 21–28 (2015).
28. Ibrahim, A. & Lin, Y. S. Pervaporation separation of organic mixtures by MOF-5 membranes, *Ind. Eng. Chem. Res.* **55**, 8652 (2016).
29. Pan, Y. C., Li, T., Lestardi, G. & Lai, Z. P. Effective separation of propylene/ propane binary mixtures by ZIF-8 membranes, *J. Membr. Sci.* **390–391**, 93–98 (2012).
30. Liu, D. F., Ma, X. I., Xi, H. X. & Lin, Y. S. Gas transport properties and propylene/propane separation characteristics of ZIF-8 membranes. *J. Membr. Sci.* **451**, 85–93 (2014).

31. Zhang, H., Liu, D., Yao, Y., Zhang, B. & Lin, Y. S. Stability of ZIF-8 membranes and crystalline powders in water at room temperature. *J. Membr. Sci.* **485**, 103 (2015).
32. Zhang, H., James, J., Zhao, M., Yao, Y., Zhang, Y., Zhang, B. & Lin, Y. S. Improving hydrostability of ZIF-8 membranes via surface ligand exchange. *J. Membr. Sci.* **532**, 1 (2017).
33. James, J. B. & Lin, Y. S. Thermal stability of ZIF-8 membranes for gas separations. *J. Membr. Sci.* **532**, 9 (2017).
34. Huang, A. S., Dou, W. & Caro, J. Steam stable zeolitic imidazolate framework ZIF-90 membrane with hydrogen selectivity through covalent functionalization. *J Am. Chem. Soc.* **132**, 15562–15564 (2010).
35. Carreon, M. A., Li, S., Falconer, J. L. & Noble, R. D. SAPO-34 seeds and membranes prepared using multiple templates. *Advanced Materials* **20**, 729 (2008).
36. Carreon, M. A., Li, S., Falconer, J. L. & Noble, R. D. Alumina supported SAPO-34 membranes for CO_2/CH_4 separation. *Journal of The Chemical American Society* **130**, 5412 (2008).
37. Wu, T., Feng, X., Elsaidi, S. K., Thallapally, P. K. & Carreon, M. A. Zeolitic Imidazolate Framework-8 (ZIF-8) membranes for Kr/Xe separation, *Industrial & Engineering Chemistry Research* **56**, 1682 (2017).
38. Wu, T., Lucero, J., Zong, Z., Elsaidi, S. K., Thallapally, P. K. & Carreon, M. A. Microporous crystalline membranes for Kr/Xe separation: Comparison between AlPO-18, SAPO-34, and ZIF-8, *ACS Applied Nano Materials* **1**, 463–470 (2018).
39. Hendlund, J., Jareman, F., Bons, A. J. & Anthonis, M. A masking technique for high quality MFI membrane. *J. Memb. Sci.* **222**, 163–179 (2003).
40. Hedlund, J., Sterte, J., Anthonis, M., Bons, A. J., Carstensen, B., Corcoran, N., Cox, D., Deckman, H., De Gijnst, W., De Moor, P. P., Lai, F., McHenry, J., Mortier, W., Reynoso, J. & Peters, J. High-flux MFI membrane. *Micropor. Mesopor. Mater.* **52**, 179–189 (2002).
41. Yu, M., Falconer, J. L. & Noble, R. D. Characterizing nonzeolitic pores in MFI membranes. *Ind. Eng. Chem. Res.* **47**, 3943–3948 (2008).
42. Yu, M., Falconer, J. L. & Noble, R. D. Characterizing non-zeolitic pore volume n zeolite membranes by temperature-programmed desorption. *Micropor. Mesopor. Mater.* **113**, 224–230 (2008).
43. Sholl, D. S. & Lively, R. P. Seven chemical separations to change the world. *Nature Materials* **532**, 435 (2016).
44. Carreon, M. A. Porous crystals as membranes. *Science* **367**, 6478, 624–625 (2020).

Index

Lightning Source UK Ltd.
Milton Keynes UK
UKHW020815290720
367351UK00001B/33

9 781786 346728